T0222447

Technical Communication Quarterly

Volume 15, Number 1   Winter 2006

TCQ Reviewers

TCQ relies on the expertise of our reviewers not only to select articles for publication but also to help authors see possibilities for development. We thank them for their contributions to the quality of the journal.

Editor's Introduction

Mark Zachry
Utah State University

As readers will note with this issue, Charlotte Thralls completed her tenure as co-editor of *Technical Communication Quarterly* with the Fall 2005 issue of the journal. During her two years as co-editor, her contributions to *TCQ* were significant. I will dearly miss our frequent conversations about the journal, ranging from discussions of mundane details to long-range plans for the publication.

From the time that Charie and I assumed editorship of the journal, we planned to work as co-editors for two years at which time she would move on to other projects and I would serve as sole editor of the journal. Those two years have passed much too quickly as we enjoyed working together on establishing new editorial policies, assisting authors in developing their manuscripts, and talking about the field of technical communication as a whole. We were also fortunate enough to enjoy collaborating on *TCQ* development and promotion in varied venues, including in our publisher's office, in Edward R. Tufte's home, at two annual ATTW conferences, and at a journal editors' forum sponsored by the Association for Business Communication.

Anyone who has considered the recent direction of technical and professional communication scholarship will attest to the fact that Charie has repeatedly helped to advance our work as a field. Those who know her well know that she has done so with intelligence, wisdom, and grace. It has been my good fortune to count her as a mentor, colleague, and friend.

Mark Zachry

Guest Editors' Introduction:
Making the Cultural Turn

J. Blake Scott
University of Central Florida

Bernadette Longo
University of Minnesota

In their influential essay "The Social Perspective and Professional Communication," Nancy Roundy Blyler and Charlotte Thralls (1993) mapped the then-well-underway social turn of the field, in the process clarifying distinctions among the major strands of social approaches. Although they accounted for the heterogeneity of these approaches, Blyler and Thralls also acknowledged the dominance of the social constructionist approach, which seeks to better understand social contexts and dynamics that specific communities deploy "in order to facilitate enculturation" (Thralls, 2004, p.125) through communicative, knowledge-building practices. In the years since the publication of "The Social Perspective," the field has continued to be dominated by this version of social constructionism, prompting us and other scholars to critique its sometimes narrow contextual focus on discrete organizational discourse communities, its mostly explanatory and pragmatic stance, and its elision of the politics of knowledge legitimation common to scholarship taking this approach (see Blyler, 1998; Herndl, 1993; Longo, 1998; Scott, Longo, & Wills, in press; Sullivan, 1990).

At the same time, a growing body of scholarship responding to these critiques has made it possible to mark what we see as a cultural turn that both complicates and extends the social perspective. Like the social perspective, the cultural one is certainly heterogeneous, drawing on a number of theoretical and (inter)disciplinary traditions and engaging in a diverse array of political projects. We can nevertheless point to several impulses that characterize this work as a whole. Such impulses are brilliantly reflected in this special issue's five essays, which account for a broader cluster of historical and cultural forces (both material and ideological) that shape and are shaped by technical communication. The essays critique technical communication's sociopolitical functions and effects, including how they help shape people's subjectivities and material experiences, and they strategize ways to intervene in and transform hegemonic forms of power. Our articulation of these impulses clearly builds on what Blyler and Thralls (1993) called the ideological branch of the social perspective, and so it would be a mistake to view the cultural turn as a radical shift from rather than an extension of the social turn. Although we are hopeful that the following group of essays will further open up the field in exciting ways, we do not present them so much as groundbreaking, radical departures than as provocative, productive extensions. In this way, the lineage that this special issue draws on includes not only cultural theorists such as Marx, Foucault, de Certeau, and Appadurai, but also sociocultural critics in our field such as Blyler (1998), Herndl (1993), Katz (1992), and Sullivan (1990).

The essays in this special issue extend the field's critical scholarship in several ways: expanding methods for talking about the influences of sociocultural contexts; foregrounding new critical perspectives on intercultural communication, safety communication, and usability; and illustrating modes of resistance and subject formation developed in extra- and counterinstitutional contexts. In "Back to Basics," Mark Longaker challenges us to read the sociocontexts of technical communication through a macroeconomic lens that accounts for historical "waves of capitalism" "driven by increased productivity, overproduction, exploitative retrenchment, and technological innovation" (p. 10). Longaker clarifies that such a lens is not necessarily deterministic but can reveal how "responses to macroeconomic developments can affect what becomes of a given crisis of overproduction" (p. 10). Considering our work within a macroeconomic context can thereby help technical communicators better strategize political action, whether focused on the disempowering effects of some approaches to usability or on the disempowering roles of technical communicators as flexible service workers for transnational corporations.

Peter Hunsinger and Jason Palmeri in their respective essays offer more targeted applications of critical cultural theory. Whereas Longaker broadens the diachronic context of technical communication, Hunsinger points us to Arjun Appadurai's theory of intertextual fluidities to broaden technical communication's synchronic, intercultural context. Although international and intercultural technical communication have becoming increasingly common areas of scholarship, Hunsinger adeptly argues that this scholarship typically works from essentialized, fixed notions of culture and cultural identity, treating the former as a "prediscursive, effectively autonomous essence posing as a set of durable habits and practices" and the latter as "something brought to communication rather than constructed and mobilized during [it]" (p. 34). In contrast, Appadurai recasted the cultural as a process shaped by a "confluence of mobile and shifting streams of textuality" (p. 39) that correspond to global movements of people, media, technologies, capital, and political ideologies. Through his explanation and application of Appadurai's work, Hunsinger demonstrates how cultural theories of globalization can serve as useful frameworks for recognizing and responding to the shifting, intertextual flows of power embodied in technical communication and the cultural identities created by this power.

Like Hunsinger, Palmeri alerts us to the potential for extending our perspective through an emergent body of scholarship, in this case disability studies. Palmeri illustrates how disability studies can help us to "usefully extend current critiques of the normalizing practices of technical communication" through readings of existing scholarship on safety communication and usability (p. 50). In the spirit of cultural studies, Palmeri moves from these productive readings to strategies for, in his terms, "disrupting normalcy" through pedagogical interventions in our teaching of technology, usability, linguistic bias, narrative, and discourse communities. Palmeri argues that the consideration of (dis)ability should infuse the full range of our research and teaching practices rather than remain relegated to special topics such as Web accessibility. The concepts of access and ease of use are, after all, at the heart of how we define technical communication and the role of the technical communicator.

The final two essays employ critical cultural theory to more specifically demonstrate and analyze the transformative potential of extra-institutional and counter-hegemonic technical communication practices. In "Cars, Culture, and Tactical Technical Communication," Miles Kimball examines two cases of extra-institutional, documentary cultures formed around automobile enthusiast publications. Kimball bases his analysis on de Certeau's distinction between institutional strategies and individual, tactical reappropriations He then demonstrates how the extra-institutional communities formed around extracorporate automobile manuals enable individuals to share and collectively build more empowering technological narratives, as well as enabling them to generate local tactical advice, new genres of Internet-based documentation, and makeshift technological artifacts. Kimball's study thus explains how users of tactical technical communication can also function as designers, "recreating technology for their own purposes" (p. 68) and even contributing to "wider cultural narratives of tactical resistance to authority" (p. 74).

Just as Kimball illustrates how automobile enthusiasts can work around and reappropriate institutional strategies designed to limit individual action, Amy Koerber illustrates how women can resist hegemonic discourse or disciplinary rhetoric that limits how they should breastfeed and even specifies what their bodies can do. Koerber draws on Foucault to read the communications and actions of breastfeeding advocates and the women they consult as different forms of resistance to disciplinary power. Some of this resistance involves negotiating among competing advice or medical guidelines set forth in regulatory discourses. Other forms of resistance entirely disrupt this regulatory framework through bodily responses outside of the framework's intelligibility. Through her rich analysis, Koerber extends our field's understanding of rhetorical agency and provides a repertoire of (un)written tactics for resistance.

In a retrospective reflection on her and Blyler's (1993) "The Social Perspective" essay, Thralls (2004) proposed a new mapping of "what the social (cultural) perspective looks like today" (p. 125). We hope that this special issue contributes to such a project, both by offering bold examples of this evolving work and by further charting a larger trajectory for theory, research, instruction, and practice. These articles begin by retheorizing technical communication practices, thereby showing theory to be, in Hall's (1996) words, a "set of contested, localized, conjunctural knowledges which have to be debated in a dialogical way" (p. 275). But they do not stop there. The authors' use of critical cultural theory enables them to build knowledge with clear implications for research, teaching, and practice, thus extending concepts of context, illuminating alternative perspectives on usability practices, setting out new international communication strategies, and drawing implications from tactical uses of extra-institutional documentation genres. More specifically, the following articles show not only how we might rethink dominant technical communication practices, but also how we can effectively resist the disempowering aspects of these practices.

In his reinterpretation of Gramsci's notion of the organic intellectual, Hall (1996) called on cultural critics to work simultaneously on two fronts: on the cutting edge of theorizing and in the realm of practice, where this theory can be applied in specific political projects (p. 268). The work in this special issue lives up to Hall's challenge and invites others to join the knowledge-building trajectory of the cultural turn.

ACKNOWLEDGMENTS

First, we would like to thank the scholars who submitted proposals and manuscripts that did not lead to articles in this special issue. The numerous outstanding submissions that we received enriched our perspective of the field and confirmed our sense that the intersection of technical communication and critical cultural studies is an increasingly busy one. We look forward to reading more about these studies in other forums.

Second, we would like to thank *TCQ* special issues editors Sherry Burgus Little and Richard Johnson-Sheehan for their assistance with the call for papers, *TCQ* editors Mark Zachry and Charlotte Thralls for their guidance, and *TCQ* managing editor Philip Parisi for his able copyediting.

Last but not least, we want to thank the friends and colleagues who strengthened this special issue with their incisive and helpful reviews: Melody Bowdon (University of Central Florida), Elizabeth Britt (Northeastern University), Michelle Eble (East Carolina University), Cheryl Geisler (Rensselaer Polytechnic Institute), Jeffrey Grabill (Michigan State University), Barbara Heifferon (Clemson University), Brent Henze (East Carolina University), Jim Henry (University of Hawaii), Carl Herndl (Iowa State University), Robert Johnson (Michigan Technological University), Johndan Johnson-Eilola (Clarkson University), John Frederick Reynolds (City College of New York, CUNY), Stuart Selber (Penn State University), Clay Spinuzzi (University of Texas at Austin), Julie Vedder (West Virginia University), and Kristin Woolever (Antioch University, Seattle).

REFERENCES

Blyler, N.R. (1998). Taking a political turn: The critical perspective and research in professional communication. *Technical Communication Quarterly, 7,* 33–52.

Blyler, N. R., & Thralls, C. (1993). The social perspective and professional communication. In N. R. Blyler & C. Thralls (Eds.), *Professional communication: The social perspective* (pp. 3–34). Newbury Park, CA: Sage.

Hall, S. (1996). Cultural studies and its theoretical legacies. In D. Morley & K.-H. Chen (Eds.), *Stuart Hall: Critical dialogues in cultural studies* (pp. 262–275). London: Routledge & Kegan Paul.

Herndl, C. G. (1993). Teaching discourse and reproducing culture: A critique of research and pedagogy in professional and non-academic writing. *College Composition and Communication, 44,* 349–363.

Katz, S. B. (1992). The ethic of expediency: Classical rhetoric, technology, and the Holocaust. *College English, 54,* 255–275.

Longo, B. (1998). An approach for applying cultural study theory to technical writing research. *Technical Communication Quarterly, 7,* 53–73.

Scott, J. B., Longo, B., & Wills, K. V. (in press). Introduction: Why cultural studies? In J. B. Scott, B. Longo, & K. V. Wills (Eds.), *Critical power tools: Technical communication and cultural studies.* Albany: State University of New York Press.

Sullivan, D. L. (1990). Political-ethical implications of defining technical communication as a practice. *Journal of Advanced Composition, 10,* 375–386.

Thralls, C. (2004). Reflection on "The social perspective and professional communication." In J. Johnson-Eilola & S. A. Selber (Eds.), *Central works in technical communication* (pp.124–125). New York: Oxford.

J. Blake Scott is Associate Professor of English at the University of Central Florida, where he teaches courses in technical communication, rhetoric and composition, and the rhetorics of science and technology. He is the author of *Risky Rhetoric: AIDS and the Cultural Practices of HIV Testing* (Southern Illinois, 2003) and, with Melody Bowdon, of *Service-Learning in Technical and Professional Communication* (Longman, 2003). Scott, Bernadette Longo, and Katherine V. Wills are the editors of the collection *Critical Power Tools: Technical Communication and Cultural Studies,* forthcoming from SUNY Press.

Bernadette Longo is Associate Professor of Rhetoric at the University of Minnesota. She is the author of the extended cultural study *Spurious Coin: A History of Science, Management, and Technical Writing* (SUNY, 2000). Her current research deals with robots, brains, metaphor, and computer history. Before her academic career, Longo was a contract technical writer for nearly 20 years in the medical and poultry-processing industries.

Back to Basics:
An Apology for Economism
in Technical Writing Scholarship

Mark Garrett Longaker
University of Texas at Austin

An economistic version of cultural studies is important to technical writing scholarship presently because capitalism's broad trends find manifestation in and are affected by local practices like scientific and professional communication. By examining their own field against the backdrop of macroeconomic eras and pressures, technical writing theorists can obtain a better understanding of the sociocultural context in which their discipline is situated, and they can better map methods of effective political action for technical communicators.

Austin, Texas is the ulcerated belly of the new economy. These are its open sores. Late one night in February 2004, in a dingy kitchen, information technology (IT) workers, denizens of the new economy, learned that their rank does not ensure against capitalism's cruelty. They drank imported beer and complained about offshoring. India and its low-paid software engineers. Their gruff bursts resonated panic as much as hate. The kitchen was lit by florescent light, the conversation peppered with ethnocentrist and isolationist platitudes. One said, "These people take American jobs." Another, "They can barely speak English!" (author's recollection based on personal communication, February 2004).

One spring night, at a bar on Sixth Street in Austin, a database manager worried over her future. She was under contract, and, once the work finished, there was none forthcoming. The jobless recovery had her angry at Bill Clinton. She said he had mismanaged the tech boom. He had left her hanging on the edge of unemployment's precipice (author's recollection based on personal communication, May 2004).

Looking at Austin, one has to wonder how all this came to be. How did technology workers, once so enthusiastic, become so bitter? How did irrational exuberance about telecommunications and the Internet turn into resentment? Why did these symbolic-analytic workers join the protectionist legions?

Cultural studies theory has appealed to those asking similar questions about complex social developments. To scholars interested in technical and scientific writing, cultural studies offers both a method for untangling history's thicket and a way to navigate it. It helps us to determine what we are dealing with and how we can respond. In its most recent manifestations, however, particularly in technical writing scholarship, cultural studies has lost the capacity to address the broad developments or the driving forces of capitalism. Viewed through the analytic schema offered by recent cultural studies scholarship, disillusioned Austinite IT workers appear to be affected by racist ideologies, and perhaps gendered language games, but they are not victims of the capitalist imperative to squeeze surplus value from labor. In cultural studies generally and in technical writing scholarship particularly, this abandonment of economism has left us unable to address one of our most influential and questionable social forces. We cannot talk about capitalism, not in any systematic way. Efforts at developing strategies to resist dominant and asymmetrical power structures will therefore fail to recognize, much less address, contemporary materiality.

Because science and technology are always tied to broader events, those interested in cultural studies and technical writing should be interested in understanding how capitalism has affected their discipline. They should be curious about capitalism's various stages, its motor forces, and its interaction with cultural discourses like the rhetoric of feasibility studies or cost-benefits analyses. Unfortunately, the variation of cultural studies most often adopted in technical writing scholarship either ignores economism or adopts its weak form. If economics appears, it usually looks like a historically isolated force, peculiar to the discrete moment and place of analysis, not driven by fundamental tendencies, not changeable across time.

This article proposes that a strong version of economism is necessary to answer the questions that cultural studies asks about technical and scientific communication: How do broad sociocultural circumstances affect this cultural practice, and how can actors intervene? Economism in this discussion references the belief that certain dynamics are central—even foundational—to a given political economy and shape its development across history. Discernible periods of capitalist development result from the repeated cycle of growth driven by increased productivity, overproduction, exploitative retrenchment, and technological innovation. A macroeconomic perspective reveals these trends, their repeated constitution, and the variations in each cycle. A macroeconomic perspective also reveals that sociocultural institutions, like manners of technical writing, often develop in dialectical response to capitalism's cycles. Though fundamental dynamics shape the broad contours of an historic era, they by no means determine what will happen in a moment of development or in the future. Sociocultural responses to macroeconomic developments can affect what becomes of a given crisis of overproduction, a given effort at technological innovation to increase productivity.

Economism, surely, has often led to deterministic theories of history, theories in which one dynamic, one institution, one cause appears to lie at the root of everything else. In its crudest formulation, this leads to the belief that capitalism causes culture, and anyone interested in generating some historical agency must grab history by the root, in this case by the mode of production. Cultural work looks like an ineffective treatment of symptoms and not the disease, the effects and not the determining cause. Cultural studies has consistently opposed strong deterministic theories of history, arguing that no single factor alone causes everything else, though some factors might have greater impact than others. In fact, cultural studies theory was developed in response to a Marxian determinism that positions the cultural worker as impotent and unable to address basic economic motivators. For the cultural studies theorist, history always appears to be a complicated arrangement, a nexus of forces, a house of cards. Absent the belief that economic developments determine the shape of human culture, we can see that cultural work has important effects.

However, economism does not necessarily lead us to a strong determinism. The economism proposed in this article does not present the cultural worker as so inconsequential an actor. In fact, by understanding the underlying economic forces that have demarcated and shaped periods of capitalist development, technical writing practitioners, theorists, and teachers can more effectively intervene in a specific moment. A broad view of macroeconomic development and capitalism's foundational role in historical periodization can lend greater agency to individuals by allowing them a glimpse at the field on which they play. Particularly, we need a more developed understanding of capitalism's stages and how they affect and are affected by techno-scientific communication. We also need to locate the culprits of social injustice. Although institutions like racism, (hetero)sexism, nationalism, and ethnocentrism certainly shape the world we inhabit, so does capitalism, and it is often at the root of our troubles. Disgruntled Austinite IT workers play a racist game, but they do so in response to economic pressure, imposed by a common capitalist drive to reduce the expense of variable capital in production, to reduce the cost of labor. If they cannot talk about capitalism, they will talk about Bill Clinton. Scholars in technical and scientific communication should not fall into similar confusions.

CULTURAL STUDIES: ITS PURPOSES AND APPROPRIATIONS

It is widely accepted among those in the cultural studies camp, those waving the postmodern banner and those bivouacked under the red flag of traditional Marxism, that America, perhaps even all of Western or global society, has changed dramatically since the 1970s. Theories about the transition from modern grainy grey to postmodern fragmented technicolor abound, and, though one is forthcoming in this article, suffice it to say presently that most theorists will register major socioseismic activity in recent decades. For cultural studies theorists, particularly those concerned with technical communication, two questions are important: (a) Where do these changes come from? (b) How can we describe recent events in such a way as to create effective manners of dealing with them? Both questions point to the need for sociohistorical mapping. We need a cartography of possibility, an agency chart.

From its origins among British theorists, long before early rumblings signaled the late–20th-century tectonic shift, the Birmingham-school variation of cultural studies tried to offer an apparatus for tabulating social phenomena in their fullest complexity. Particularly, Williams (1977) and Thompson (1966) offered compli-cated theories about how economics interacted with cultural forces to create en-tire social formations grounded not in only the material or the social, but in both and in their interaction. Williams, for instance, developed a number of theoreti-cal terms (*structure of feeling; hegemony; dominant, residual, emergent cultures*) to discuss the contingent construction and material effects of phenomena that in traditional Marxian theory would appear "'superstructural': a realm of 'mere' ideas, beliefs, arts customs, determined by the basic material history" (p. 19). Thompson, likewise, in his investigations of class formation and 18th- and 19th-century British popular culture, developed a vocabulary to discuss cultural events, not as the consequences of economic circumstances but as participatory elements in broad and contingent developments. The English working class was not summoned by an economic conjurer but was instead produced in a brew of shared material circumstances, common practices, and political struggles, a rec-ipe to which workers contributed. The working class "was present at its own making" (Thompson, 1966, p. 9). At the beginning of cultural studies theory, there appeared radical potential for those interested in cultural events and in the agency of cultural workers.

But neither Williams (1977) nor Thompson (1966) abandoned the notion of de-termination. They preserved an economism while avoiding its excesses and while incorporating the tools of cultural analysis. Williams said that "Marxism without some concept of determination is, in effect, worthless" (p. 83). To escape the trou-bling implications of an abstract sense of determination, Williams examined peo-ple's accessible, inherited, and contingent conditions. Determination becomes a set of "limits and pressures," "complex and interrelated," and society becomes a "constitutive process" (pp. 85–87). Capitalism's driving forces were never forgot-ten, but cultural institutions received attention as influential moments in a social formation and viable sites of resistance. Then there was the 1970s.

The Birmingham Centre for Cultural Studies entered a moment of dramatic social and historical change armed with a graphic apparatus for diagramming the interplay of cultural factors, received material conditions, and human agency.

Like most critical schools surviving this era, the Centre left a tattered reminder of its former self. Particularly lost from its developed corpus was a regard for economics. Though the major thinkers in the second generation believed that Western societies were undergoing a sea change, and though many alluded to material forces (capitalism, industrialism, managerialism), lacking a developed vocabulary for discussing macroeconomics, the material strength of this theoretic body atrophied while the cultural analytic muscles developed. Cultural studies became truly cultural. This is not to say that all of cultural studies theory lost track of economics. Jameson (1991), often grouped in the cultural studies genus, clearly focused on capitalism's foundational role in historical and cultural development. Nevertheless, the dominant version of cultural studies theory coming from the Birmingham Centre did turn its emphasis away from the notion of economic determinism.

The two major Birmingham-influenced analyses of postmodern society and politics—one by Hall (1988) about Great Britain under Margaret Thatcher and the other by Grossberg (1992) about the United States under Ronald Reagan—either avoided or derided discussion of economic determination, even if done in a qualified manner. Hall claimed to preserve some sense of economic causation, that is, of Thatcherism's design to expand the prerogatives of capitalism (p. 4). He professed some awareness of Britain's need to stabilize political order in "rapidly changing global and national conditions on an extremely weak, post-imperial economic base" (p. 30). He even made veiled references to the decay of structural economic conditions and a crisis of overaccumulation (pp. 43–44). Yet Hall focused his analysis on cultural events such as the moral panic invoked by Thatcherite conservatives, their authoritarian populist rhetoric, and the cruel abuses of civil rights, which was a brutal retrenchment shielded by law and order's auspices. At one point, Hall derided economistic invocations of "an abstracted 'tendency of the rate of profit to fall'" while, in the same breath, he reduced the leftist project to a cultural struggle for increased democracy (pp. 124–127). In the United States, Grossberg likewise adverted to economic causes while also emphasizing cultural manifestations and solutions. Though Grossberg discussed various stages in capitalist development, even mentioning new forms of accumulation as if they were functional components in the shift to postmodern society (pp. 347, 355), he said that the new conservatism is "put into place through cultural rather than political struggles" (p. 15).

Surely, Hall (1988) and Grossberg (1992), like Williams (1977) and Thompson (1966), had good reason to step away from the notion that economic motivators trump all cultural institutions. If determinism means that economic forces cause every detail of cultural development, then cultural workers can do nothing but inhabit the epiphenomenal discourses that (falsely) reflect capitalism's basic motor functions. And resistance will be limited strictly to complete overhaul of the economic base. But Williams demonstrated that determination need not be so totalizing a concept. In fact, keeping a revised notion of economic determination in view can reveal aspects of our historical situation, a recognition necessary to any viable effort at resisting economic institutions through cultural ones. Among the second-generation Birmingham thinkers, however, even this revised version of determination began to wither. After the Birmingham Centre's main figures stepped away from economism, the doors of cultural studies theory opened to a variety of characters, hailing from sociology and anthropology to critical philosophy. In this decidedly cultural phase, technical writing scholarship discovered cultural studies.

Culture's hypertrophy is illustrated clearly in several recent efforts to analyze technical writing pedagogy and practice. Herndl's (1993) critique of dominant trends in technical writing pedagogy from a cultural studies perspective, for instance, invoked "Marxist theories of culture" while sidestepping "the pessimistic and mechanical model of social reproduction" (pp. 351–352). Herndl's analysis relied heavily on sociological theories of agency adopted from Anthony Giddens, and, though there was reference to "material" practices, they were not set in broad economic context (1993, p. 354). In Herndl's lexicon, the *material* references the immediate circumstances of labor, not the broad trends or abstracted forms of capitalist exploitation. Likewise, Scott's (2003) more recent application of a cultural studies approach to home HIV-testing apparatuses and literature located scientific production within the "broader social conditions of possibility ... map[ping] the connections and power relations among science's heterogeneous actors" (pp. 354–355). Scott, like Herndl, considered "material circumstances" (p. 356) but did not include capitalism's diachronic and global economic conditions under this umbrella. Instead, his analysis of Confide's FDA approval and sale focused its economic lenses on Johnson and Johnson's considerable financial muscle, its marketing campaign, the corporation's donations to prominent politicians who influenced the product's approval, and the broader 1990s effort to privatize and marketize health care (pp. 359–660). What resulted was an admirable and complicated analysis, which certainly considered economic factors often overlooked in studies of scientific development, but which was unable to locate Confide's relation to larger trends in U.S. capitalism.

Connections between the material influences that Scott (2003) noticed in his analysis of Confide and a broader economic trajectory have been made in recent Marxian analyses of the biopolitical nature of postindustrial capitalism. Hardt and Negri (2000), pointing to the "real subsumption" of everything under capitalism's domain, claimed that postindustrial dominance of immaterial labor imposes market forces on once quasi-sheltered institutions such as health care (pp. 22–41). To put this in more accessible terms, capitalism in the 1970s stopped being so strictly concerned with producing things in the factory and began finding ways to produce people in their homes, in their public spaces (movie theaters, malls, gyms), even in their private enclaves. Hardt and Negri noticed an advanced manifestation of a trend that Braverman (1974/1988) called "the universal market." Since its inception in the 18th century, Braverman noted, capitalism has consistently found its way deeper into people's lives, inserting the mechanism of monetary exchange wherever a breach appears (pp. 188–196). Confide, though certainly not an effect of capitalism's unbending causal forces, does appear to be a locally produced manifestation of the market's ever-growing purview.

Though Hardt's and Negri's (2000) analysis is quite vague, and though their invocation of the Foucauldian notion of biopower sometimes appears abstract to a fault, their point is salient nonetheless; health care is tied to developing economic institutions. The global market for prescription drugs, corporate efforts to enforce intellectual property rights across national boundaries, the transportation of human bodies to nation states where treatment is cheaper (and often riskier)—these recent developments comprise a global system swallowing every aspect of human life, occupying every corner of physical existence, moving bodies and technologies along marketized conduits. Whether we call it biopower or the universal market, we are still looking at a fundamental economic drive to expand the arenas of accumulation. This is not to say that Scott's (2003) analysis is wrong or even faulted, only this: In his efforts to look at social circumstances, Scott did not open the lens to its widest possible aperture.

Others melding cultural studies and technical writing have produced similar results. Longo (2000), for instance, opened her history of technical writing with a professed desire to consider the expansive social field and its effects on the profession's development. Longo's critique of previous scholarship echoed Herndl (1993) and Scott (2003). She contended that theorists and historians have decontextualized the practices of technical communication, thus failing to understand the conditions and effects of knowledge production and therefore also failing to imagine interventions into more expansive social conditions such as epistemic authority, disciplinary hierarchy, or professional formation (pp. 7–18). Longo's solution to this shortcoming was a cultural studies approach to the formation of technical writing, extending back to the 17th century and including important developments in capitalism such as the 19th- and early 20th-century creation of the vertically integrated corporation and the culture of managerialism. Longo's detailed account of the concomitant and interrelated appearance of technical writing and managerial capitalism made no mention of the macroeconomic forces underpinning this stage of historical development.

Economists writing during managerialism's heyday, however, did locate material roots to the phenomena that Longo (2000) recognized as important to technical writing. The large corporation and its culture of efficiency began, as Baran and Sweezy (1966) argued, with a break from the entrepreneurial model, which exposed individual fortunes to frequent economic downturns and cutthroat competition. To preserve the capacity for accumulation without suffering the cruel discipline of an untamed market and to facilitate large-scale mass production, corporations integrated, grew large, chose management structures over individual directors and shareholders over sole proprietors (Baran & Sweezy, pp. 14–52). With market forces held in check, prices behave more as they do in a monopoly system and control becomes the guiding principle of everything from technological innovation to workforce management. In these conditions, growth is steady and the surplus rises, which, without safety valves, could lead to a crisis of overproduction (pp. 67–72). Baran and Sweezy identified numerous institutions developed in the 20th century to absorb surplus capital—among them the military–industrial complex (pp. 178–217)—whose disciplinary culture impacted technical writing and whose undemocratic strictures have garnered much attention among technical writing scholars (Sullivan, 1990). If Baran and Sweezy were right, then Longo's history, though certainly astute and well researched, overlooked the economic forces contributing to technical writing's development.

The scope of Herndl's (1993), Longo's (2000), and Scott's (2003) analyses has been repeated in a host of scholarship at the intersection of rhetoric and cultural studies generally, and technical writing and cultural studies specifically; noticing that this work to date has not dealt with economic issues, however, is no reason to insist that such a perspective be adopted. Surely, Grossberg (1992) and Hall (1988), addressing expansive moments of conjuncture, should have paid greater heed to capitalism, but why should technical writing scholars care about the economic underbelly of history's lumbering progression? The answer to this question has already been provided by Herndl, Longo, and Scott: Social circumstances affect and are affected by various cultural networks and forces. More important, intervention into history requires a full understanding of its machinations. Herndl, Longo, and Scott hoped that a more developed system of mapping the social field would allow actors to play on and to alter its topography. This consideration of how economy and culture interact, though rare, is not entirely absent from the body of technical writing scholarship. It is exemplified by Henry's (2001) efforts to closely analyze students' working conditions through postindustrial autoethnographies. Henry himself noticed that there has been little work done on "workplace literacy demands as shaped by the early twentieth-century U.S. economy and its modes of production" (para. 2).

Of course, attention to capitalism does not mean ignorance of the more identifiably cultural events mentioned so far. The challenge is to consider the fulcra of economic pressures and cultural institutions in the interest of progressive social change. Though this article cannot present a full explanation of economism's utility, it can offer a brief sketch of what economism can add to our understanding of technical communication by presenting an outline of a theory of capitalist development, two important capitalist eras (the industrial and the postindustrial), and the implication of technical writing trends in these eras.

LONG WAVES OF CAPITALISM:
A THEORY OF HISTORY INTO THE PRESENT

Traditional Marxian economics of the early 20th century posited that capitalism would collapse necessarily of its own systemic crises, a thesis that seemed entirely reasonable given late–19th-century economic volatility and the 1870s deep global depression. In this period, however, a Soviet economist, Nikolai Kondratiev, argued that capitalism had within itself a mechanism of rejuvenation through concentration and investment of liquid capital into technological innovation (see Schumpeter, 1939/1982, and Mandel, 1972/1978, for excellent summaries of Kondratiev's theories and the implication of techno-scientific innovation in long capitalist cycles). Kondratiev argued that since the 18th century, global capital has endured 50- to 60-year cycles, each beginning with a revolutionary change in the means of production, followed by a 20- to 30-year period of growth during which labor can make increasing demands of capital, punctuated by a crisis of overaccumulation, and then ending with a 20- to 30-year steady decline in which profit margins narrow while capital is increasingly siphoned from wages and physical plant. The middle of each cycle brings financial crisis and begins a steady concentration of liquid capital into investment that eventually produces some innovation, thereby making profit realizable in large part by increasing productivity. In short, when a combination of technological parity among producers and overaccumulation makes profits difficult to realize, capital does two things: It turns the screws on labor, and it begins looking for new productive avenues. These two responses correspond to Marx's (1867/1976) division between absolute and relative surplus value (Harvey, 1990, p. 186). When the market will not yield profit, the capitalist can lengthen the workday (increasing absolute surplus value) or look for ways to increase productivity (increasing relative surplus value; Marx, 1867/1976, p. 432). In fact, during moments of global crisis, both actions are necessary. The resources taken from workers allow investment into and eventual production of innovation, which increases productivity. More recent economists working from Kondratiev's hypothesis have divided history into four (possibly five) long waves, each unique in its realization but following a common pattern (see Table 1).

Though the aforementioned description relies on Marxian economists, it should be noted that Kondratiev's theories have influenced a number of capitalism's cheerleaders, most notably Schumpeter (1939/1982), who accepted the theory of long waves, further claiming that competitive disequilibrium is necessary to innovation in the second half of each cycle (pp. 73–103). Schumpeter's (1942) term for the place of technological innovation in capitalist rejuvenation, *creative destruction* (pp. 81–86), now hangs on the lips of nearly every NASDAQ acolyte who kneels before the Internet alter, whispering prayers for financial regeneration. Recently, economists Abramovitz and David (2000) developed a complicated method

TABLE 1
Long Waves of Capitalist Production

Era	Expansion	Contraction	Means of Production
1. Pre-Industrial	1793–1825: Steady growth and economic success allowing the Federalist era and the era of good feeling in American politics	1826–1847: Crisis (including the famous banking crises) and tensions associated with Jackson-era politics	Artisan-produced machines, local manufacture, plantation-style agriculture
2. Early Industrial	1848–1873: Economic successes associated with the railroad and the telegraph industries allow westward expansion of the United States and growth of markets	1874–93: Deep depression and a series of recessions lead to labor-management strife, progressive-era reforms, and the first efforts to unionize on a broad scale	Machine-produced machines, concentration of manufacture in the factory, railroad transportation, and telegraph communication
3. Late Industrial	1894–1913: Widespread prosperity allows for euphoria preceding World War I	1914–1939: Two world wars and a series of financial crises cripple the world and the U.S. economies	Further expansion of machine-produced machines into industries beyond the railroad, innovations in the chemical industry (synthetic materials like rubber, nylon), electricity, and widespread availability of telephone
4. (Post) Modern	1940–1966: The "golden era" of American capitalism allows for a growing middle class and a uniform sense of prosperity	1967–1995(?): Recession and stagflation of the 1970s followed by "hard" decisions to bust unions and control inflation in the 1980s	Further expansion of the factory mechanism, increased armament production, introduction and common use of computing technology, and air travel
5. Fast Capitalism	1996–Present: Possibly a new expansion in productivity spurred by information technology		Telecommunications, advanced statistical applications, and monitoring technologies

Note. Loosely adapted from Mandel, 1978, pp. 130–131.

of measuring productivity growth (total-factor productivity) across history and arrived at a periodization very similar to the one presented earlier. More remarkably, Abramovitz and David positioned technological innovation as one of the main drivers behind long waves of economic growth (p. 41). Alcaly (2003), an economist and hedge-fund manager, has become particularly fond of Abramovitz's and David's belief that innovation in the latter part of the 20th century has led to a capital-shallowing movement away from investment in tangible assets and toward information (Alcaly, 2003, pp. 44, 62). Alcaly believes that in the late 1990s, the U.S. economy entered the upswing of another long wave (pp. 35–51). These predictions may be a bit optimistic. Others have argued convincingly that the productivity miracle of the late 1990s was unremarkable in comparison to productivity growth during the early 1960s when the U.S. economy was certainly in an extended upswing (Brenner, 2002, p. 223). And this recent productivity growth, certainly overstated, might have resulted from typical efforts at expanding absolute surplus value by depressing real wages, a common trend since 1966 (Henwood, 2003, pp. 38–62, 203–204). For these reasons, we scholars might be skeptical about whether a fifth long wave has begun. Economics, though a poor predictive science, is an excellent historical science. Like rhetoric, it might not allow us to understand what will happen, but it can help us to understand what has happened, what is happening, and what we can do in the present to manage our circumstances.

In the long-wave theory of history, there certainly lurks a skulking determinism that cultural studies theory has resisted. After all, Marx (1867/1976) did proclaim that a change in the mode of production is "the necessary product of the revolution in the means of production" (p. 602), a statement that has led many to accuse him not only of economism but also of technological determinism (Street, 1992, pp. 30–31). If strict economism or technological determinism have any merit, then the type of study I am proposing is politically impotent. However, one can accept long waves in capitalism's development—even accept the operation of tendencies like overaccumulation and creative destruction—without accepting any necessary results. The means of production might determine a change in the mode of production, but the nature of that change remains open. The cultural theorist's task, in this case, is to locate possibilities. Of course, this will require identifying economic pressures and their immediate connection to useful practices. In the history of technical writing practice, there have been two immediately identifiable moments when stages in the long development of capitalism made possible writing practices that were potentially exploitative and empowering. These are mid–20th-century Fordism and the late–20th-century service and information economy. A brief glance at each and at the possibilities available therein demonstrates the kind of mapping that this article promotes.

The story told in each of the sections that follow employs a necessary but oversimplified bifurcation of history's progressive and exploitative possibilities and a complementary split between resistance and complicity. Surely things are never so simple. And actors in any moment will always find themselves both resisting and complying with the power structures that they inhabit. Nevertheless, to illustrate the utility of mapping available and viable strategies in a moment of conjuncture, this analysis paints in rather stark lines what are really blurry territorial distinctions. Fuller analyses, informed by the economism advocated in this article, should take account of what Herndl (1993), adopting Giddens's vocabulary, called the *structuration* of every historical moment, the complicated construction of an entire social formation and the implication of every actor both in its reproduction and its alteration.

MID–20TH-CENTURY FORDISM AND MANAGERIAL WRITING

The mid–20th-century phenomenon shorthanded as Fordism began during the 19th century as the vertically integrated corporation, pioneered in the railroad industry, and was slowly adopted in other sectors over the course of a century. As Harvey (1990) described it, Fordism was more than mass assembly-line production, even more than vertical integration. In effect, Fordism was a whole way of life, still nascent during the financial crisis that marked the middle of the late industrial era (roughly 1913–1914; see Table 1). Fordism's rise and apotheosis allowed the modern era, also known as the golden era of American capitalism, which was the first half of the fourth long capitalist wave. Through the incorporation of various control mechanisms—scientific management of labor, city planning, global management of trade, Keynsian management of monetary supply, state regulation of industry—the Fordist mode of production staved off for a time the detrimental effects of an unfettered capitalism (Harvey, 1990, pp. 125–140). Fordism's downfall was its inflexibility, a characteristic magnified by the 1970s financial crisis that began the fourth wave's postmodern slope. When accumulation became difficult and capitalists needed to find new sources of surplus value, the welfare state, the trade union, and government regulation all became hindrances.

In technical writing, especially during the postwar years, Fordism encouraged the same kind of controlling discourse that Yates (1989) found operative at late–19th-century railroads. This was an effort at control through communication. In early genres like circular letters, manuals, notes, forms, and in-house magazines, Yates located "downward communication" that "was critical to implementing executive plans and decisions" along with "upward communication … critical to formulating them" (p. 77). By the 1920s, these genres and others, particularly reports and memos, were taught in formal professional-writing classes at American universities like New York University (Yates, 1989, pp. 92–93). At the beginning of the modern era, popular technical writing textbooks codified and taught Fordist communication strategies, efforts at "recording and reporting developed for systematized management and control" (Longo, 2000, p. 115). In a review of the profession, Gould (1964) found the most opportunities for technical writers in major corporations or at government-funded labs, where research reports had to give management sufficient information about the enterprise. Codified reports facilitated control of the research organ in an integrated corporate biology (Gould, 1964, pp. 67–68). Though Fordism may have begun in the late 19th century, its effects and its manifestations in technical writing continued well into the 20th century. Fordist technology (including management systems and physical plants) boosted productivity in two eras of capitalist development, enabling managerial rhetoric to persist for a long stretch of history. Sauer's (2003) work on communication in late–20th-century mines located a managerial rhetoric still operative in hazardous industries where worker safety is governed by federal regulation, constant corporate supervision, codification of practice, and efforts to record disaster to prevent recurrence. Structural conditions in certain industries have led to managerialism's persistence despite its shortcomings.

Recent analyses have found that managerial communication failed for the same reasons that Fordism failed: a lack of flexibility. While looking at both the Three Mile Island incident and the Shuttle Challenger disaster, Herndl, Fennell, and Miller (1991) found that the typical Fordist divisions in integrated industries created professional cultures that interpreted information differently, leading twice to miscommunication and catastrophic results (pp. 286–295, 300–303). In a study of the Chicago Transit Authority in the mid 1970s, Coogan (2002) identified similarly inflexible methods of management–labor interaction, which led to repeated communication failure and disastrous results (pp. 289–291). Sauer (2003) noticed that in contemporary mines, the scientific knowledge that gets privileged, codified, and passed down to workers is less effective in averting disaster than an individual miner's "pit sense" (pp. 81–83). The highly structured regulations drafted by the federal government and the routines developed to manage hazardous conditions at individual corporations often fail because they are not sufficiently flexible to help individual miners negotiate complex and changing circumstances. Across four separate industries, these studies point to a common failing of Fordist organization, a common shortcoming in managerial communication: an inability to adapt.

Although economic pressures surely contributed to managerial hierarchies and to inflexible communication practices, they by no means determined the uses and eventual outcomes thereof. Technical writing scholars like Sullivan (1990) are right to worry over the Fordist controlling and antidemocratic appropriation of technical communication. All the same, we scholars cannot forget that Fordism, although it reduced human freedom, expanded economic equality by raising the standard of living for many. The welfare state and the governmental regulatory apparatus are now enclosed pastures where the great American middle once grazed. Sauer (2003) argued that federal regulations—a decidedly managerial communication—were a welcome development in the mid–20th-century mining industry because, although such regulations reduced flexibility, they also made miners less dependent on corporations' good will for their own safety (p. 33).

In many cases, as Longo (2000) herself illustrated, managerially directed technical communication tends to favor capital over labor (pp. 97–99). Nevertheless, managerial communication has also been adopted to protect and even expand the rights of workers. Traditional grievance procedures embody the same inflexibility, the same penchant for codification and control that one finds in other managerially inflected documents, but they often serve a prolabor agenda. Perhaps workers did lose individual autonomy, but they gained pensions and health care plans, a fair trade by many scales. Recent corporate efforts to cut away at these institutions by replacing pension plans with individual retirement accounts, to institute more flexible working arrangements, and to heap the responsibility of health care payments onto employees (all under the aegis of greater choice) demonstrate how valuable these distinctly managerial institutions have been for workers.

THE LATE–20TH-CENTURY SERVICE ECONOMY AND USER-CENTERED DESIGN

One need not travel far nor listen long to hear proclamations against Fordism both in the business press and in technical writing scholarship. The new economy, the networked organization, the free agent nation—these rhetorical bugles summon a cavalry of corporate reformers demanding chaos, workplace democracy, and worker autonomy. After the early 1970s crisis, a radical post-Fordist restructuring of the U.S. landscape manifested itself in a number of ways: the shift of capital investment from physical plant to the financial sector; the restructuring of the labor market, including the erosion of steady work and the creation of more part-time positions, more flextime arrangements; the development of just-in-time production and yield management. These and many others are all strategies to undo Fordist rigidities and to restructure capitalism so that higher profits can be realized in the short run and so that innovation can occur in the long run, and, above all, so that a new cycle can be made possible (Harvey, 1990, pp. 141–172). As Fordism withered, the business press sang hosannas to the coming of the new organization, and technical writing scholars fell in tune. In recent articles, one finds common post-Fordist refrains to "employee empowerment, synergy" (Thompson, DeTienne, & Smart, 1995, p. 163) and to "democratic communication" (Waddell, 1995, p. 208). Just as the business press ignores the potential stresses and cruelties of the new work environment, so do technical writing scholars often overlook the effects of workplace flexibility on postindustrial laborers, technical writers particularly. Technical writers in Henry's (2001) classes regularly recorded unbearable workplace stress and instability, leading him to declare that flexibility "augers poorly for professional writers ... in many ways, for the majority of people who enter the workplace in years to come" (para. 2).

Of course, a common rhetoric and a common historical field do not add up to a common purpose. Though technical writing scholarship certainly participated in the broader shift from Fordism to post-Fordism, it was not necessarily or ever simply complicit with the profiteering agenda that dominates corporate boardrooms and cuts into technical writers' job stability. Technical writing scholars advocating typical post-Fordist cures to economic crises might find themselves on the exploitative side of history. Or they might find themselves forging a new progressive agenda out of available resources.

A number of developments in technical writing scholarship suit the post-Fordist landscape and exhibit the same multiplicity of potential located in managerial communication. The user-centered design movement, for instance, has both the potential to help realize profits in a tight commercial market and to empower consumers and workers (users) typically caught in an asymmetrical relationship with producers and/or management. The user-centered design movement began after capitalism's fourth wave entered its descending slope. In this environment, user-centered design looks like one manifestation of a larger trend: the shift to service and to value-added industries to reduce the indeterminacy of sale in an increasingly competitive and tight market. In the 1970s and 1980s, as productivity declined and technological parity among producers manifested itself, realization of profit became more difficult, and money poured into industries that focused either on raising the market's saturation level or on encouraging consumers to choose one product over another. Among the more visible symptoms of this trend are the burgeoning consumer-credit and advertising/market-research industries (Mandel, 1972/1978, pp. 397–400).

Capitalist ecumenists are fully aware of these developments, and they often sermonize about the experience economy and the age of style. For Pine and Gilmore (1999), the turn to value-added industries means an entirely different way of doing business, one that focuses less on the manufactured product and more on its related services and experiences. Whoever offers the better experiences gets the sale. Pine and Gilmore's excited declaration that the "customer is the product" (pp. 163–184) demonstrates that, in its search for favorable exchange, capitalism has moved into the production of human subjectivity and even of human bodies—biopower. Not to be outdone, Postrel (2003) giddily gossiped about postmodern capitalism's affair with style. At the margins of value, in increasingly competitive markets, producers focus on value-added industries, particularly the aesthetic qualities that differentiate similarly functional items like cars, restaurants, and even computers. In Postrel's typically ebullient prose: "In a crowded marketplace, aesthetics is often the only way to make a product stand out. ... [So] aesthetic creativity is as vital, and as indicative of economic and social progress as technological innovation" (pp. 2, 16).

Technical communicators interested in document design have adopted the service industry's vocabulary and evaluative framework. Particularly, *quality* and *value-added* appear regularly, indicating that the document-design movement developed at least in part to address contemporary economic pressures. Schriver (1993) demonstrated that in the early 1990s consumers were willing to pay more for a product accompanied by a clear instruction manual. Of those who answered *yes* and *maybe* (63.1%) to the question "Would you be willing to pay more for a product if you knew it had a clear manual?" most (79.8%) were willing to pay a significantly higher price ($3 or more) for the addition of a high-quality document (Schriver, 1993, p. 247). Schriver's concern in this study, even her comparative argument, match Pine and Gilmore's (1999) exuberant claims. In fact, Pine and Gilmore similarly compared the price of a cake's ingredients with the price customers will pay for a "birthday experience" to find, as did Schriver, that there is much greater potential for profit realization in the latter effort (p. 21). Redish's (1995) belief that technical communicators should speak "the manager's language" of added value is a more glaring example of an experience-economy vocabulary among technical communicators (p. 26).

Though a number of technical writers appropriate the lexicon, even the profiteering of late–20th-century capitalism, the design movement is not entirely resigned to such efforts. Schriver (1997) played to capitalist interests while championing a group typically on the abused side of an asymmetrical relationship: consumers. Although Schriver appealed to corporate interests with talk of quality and greater profits, she also tried to empower readers who might find technology off-putting, intimidating, or exclusionary. She showed particularly admirable concern for those most often located on the wrong side of the digital divide: the elderly and the physically challenged (1997, p. 247). Others, like Johnson (1998), have made convincing claims that user-centered design can bring an egalitarian politics to the imbalanced relationship between technology experts and users. Reconceptualizing user knowledge as a valuable practical understanding necessary to the development of technological artifacts (be they manuals or videocassette recorders) can prevent experts from imposing technologies without consulting all those affected. For Johnson, user-centered design is, therefore, a democratic project (pp. 57–67). Though Johnson and Schriver never considered the broad economic conditions that make their project possible, they did find progressive prospects in an available development.

User-centered design's dually progressive and exploitative potentialities can be demonstrated with a brief look at one of its chief research methodologies: prototyping. Originally developed by Scandanavian researchers in an effort to make the workplace more democratic, prototyping empowered unionized workers by giving them a say in the development and implementation of new production technologies and workplace platforms. In an early user-centered research program, Marxian researchers began the UTOPIA project (1981–1984) to empower workers. They built crude mock-ups of workstations and asked workers to play at these stations, simulating their work practices and their craft knowledge. This knowledge could then become an important component in the development of new workstations, computer systems designed to suit workers' knowledge and promote workers' authority. As prototyping traveled, however, it was increasingly situated to suit the prerogatives of contemporary capitalism. In some research, the political project of worker emancipation was pushed out by the ethical and professional problems of developing morally and legally acceptable systems. In the United States, prototyping was fully folded into the postmodern effort at reducing the indeterminacy of sale. Consumer-focused research added value to products by learning what experiences consumers would enjoy (Bjerknes & Bratteteig, 1995, pp. 77–89). In this last phase, dubbed *contextual design*, prototyping became "emphatically capitalist" and empowerment became "functional" not "democratic" (Spinuzzi, 2002, pp. 208–209). There is something laudable in contextual design, especially as it includes those left out of the technological loop, but there is also something lost in the transition from UTOPIAn mock-ups to Schriver's (1997) user-centered protocol analyses. The latter is certainly more in line with contemporary capitalist imperatives. Of course, without economistic lenses, this is not evident.

CONCLUSION: FROM MAPPING TO STRATEGIZING

The aforementioned vocabulary of *progressive* and *exploitative*, though certainly illustrative, flirts with oversimplification. Getting on the right side of history is never so easy, nor so easily plotted. Contextual design, though not part of a socialist project, does have democratic potential and is therefore admirable. User-centered design's multiple appropriations demonstrate that the progressive and the exploitative are always both present in a moment, an artifact, or an action. The challenge is to actuate one while suppressing another, a task that will never simply involve adopting or abandoning a practice. Actuating the progressive possibilities in a moment of conjuncture requires strategy, which must begin with a detailed map of the field.

Cultural studies appeals to technical writing scholars by providing the tools necessary for both mapping and strategizing. Cultural studies offered Scott (2003) the opportunity to get involved in the issues that he researched, the ability to understand a field of power and to intervene productively. For Herndl (1993), cultural studies offered a critical discourse, one that could describe, explain, and resist (pp. 349–350). Invoking a vocabulary not unlike the one adopted in this article, Herndl (1996) spoke of trying to avoid "the dark side of the force" (p. 455). In one study of a biologist employed at a military research facility, Herndl (1996) found admirable moments when the biologist fought the institution's culture, working tactically to reappropriate rhetorical elements made available to him (pp. 464–466). In all this discussion of resistance, one has to wonder what is being resisted. Herndl's (1996) use of cultural studies allowed him a rich vocabulary to discuss quotidian tactics, but, like much of contemporary cultural studies scholarship, these analyses lacked a broad sense of history. Scott (2003) and Longo (2000) provided more encompassing historical sketches and were therefore more able to locate targets for their political salvos, but even they could not place the crosshairs on capitalism. This is what macroeconomic analysis brings to cultural studies scholarship, to technical writing scholarship particularly. Not only can we scholars map the long history of capitalism and its particular manifestations in a field, but we can also map its trajectories, its possibilities, and new available means of resistance.

This article already touches on a few methods of resistance in particular moments of capitalist development. In the era of managerial capitalism, the AFL-CIO created hierarchical unions that, although they controlled workers, organized them in a manner sufficient to resist managerial exploitation. In the United States' service economy, user-centered design resists the exclusion of the disabled, even though this same movement plays directly into the hands of a postindustrial imperative to reduce the indeterminacy of sale. A strong version of economism also gets technical writing scholars and their students to recognize the common work that technical writers perform (the common work done by all intellectual laborers) and makes possible the beginning of a new effort at organized labor, one able to resist the cruelties that Austin's IT workers know quite well, that Henry's (2001) students recorded in their autoethnographies. Economism helps technical writing scholars, teachers, and practitioners to recognize that they share a common and hostile environment but also that they share a common labor. We are, as Henry said, "discourse workers" (para. 15–21). Identification as such is the first step toward collectively acting and resisting exploitative, unstable, and stressful working conditions. Just like Thompson's (1966) 19th-century British factory workers, discourse workers, through a strong version of economism, can more fully participate in their own class (trans)formation and can fight collectively for their rights as workers. If there is any validity to the speculation that we are presently ascending the upward slope of a new capitalist wave, then this is a prime historical moment for organization among knowledge and service workers, the two principal new-economy classes. In the 20th century, labor made its greatest gains during growth periods like the progressive era (1894–1913) and the golden era (1940–1966) of American capitalism. Even if we are not at the cusp of a newly prosperous age, we are witnessing a new privilege among the heralded symbolic-analytic workers. Presently, terms like *knowledge worker* and *the creative class* hang in the air, indicating a newly possible class consciousness not just among technical writers but among all intellectual laborers.

Without the ability to map, to name, and to describe, we in the field of technical writing will never be able effectively to act collectively. Without the ability to theorize capitalism's various stages, we will never be able reasonably to determine a viable manner of altering its shape. A blindness to capitalism's place in history (broadly) and in technical writing (specifically) can dangerously lead to the postures adopted by the Austinite IT workers whose stories began this article. Poorly understanding the situation can lead to bad policy. Present-day laborers who channel legitimate grievances through a narrow view of history and of capitalism's role therein may react rather than resist, and they may do so individually rather than collectively. In fact, failure to trace much of the postmodern world to capitalism's systemic developments has made political reaction the most recognizable and appealing response to real material suffering.

Not only can economism help us to escape political reaction, but it can also help us to form a viable and effective response to this specific moment in capitalism's development. Like some in the user-centered design movement (Spinuzzi, 2002, p. 215), Hardt and Negri (2004) argued recently that the traditional labor union is no longer a viable strategy. Instead, they advocated an appropriation of biopower's resources to create a global drive toward democracy. Their invocation of "swarm intelligence" (p. 93) and their descriptions of particular movements (in Italy, the Chiapas, Seattle) seem vague if not disjointed, yet they made an important and noticeable point. As capitalism changes, so must our responses. Perhaps the managerial union is not a useful manner of organizing discourse workers, but new models of unionization are developing daily. The Service Employees International Union organizes across an inclusive swath of professions, rejecting the AFL-CIO model of sectioning into trade unions. The Industrial Workers of the World organizes broadly without trying to occupy or close individual shops. There is even recent talk of open-source unions that recruit individuals across rather than within cohorts of geographically or institutionally bound workers. Intellectual workers very well may need new methods of organization. These are groping efforts to locate and develop strategies.

Humility is requisite in these efforts. It is a bit absurd to hope for transnational or permanent revolution in technical and scientific writing. Hardt and Negri (2004) noticed, rightly I think, that information workers will not (and should not) become a vanguard party to lead the revolution. Organizing labor in the networked society requires coordinating unionization efforts with other segments of society, other social movements (Hardt & Negri, 2004, pp. 137, 223). Like recent demands in the Philippines and the Ukraine for more democratic governance (efforts in part made possible by market liberalization and information technology), technical writing falls within capitalism's purview. Yet such moments of influence hold possibility for action. The task is to locate those moments for organization and coordination. Without a developed understanding of how broad social forces impact and are impacted by seemingly local practices, this kind of strategizing is impossible. Cultural studies provides technical writing scholars both the cognitive instruments for mapping and the tools for strategizing. More specifically, economism provides us with the wide lens needed to understand diachronic material forces and their local manifestations. Put to use, both can help us to understand and make good practice of scholarship and scientific writing.

ACKNOWLEDGMENT

Special thanks to Clay Spinuzzi, J. Blake Scott, and to the *TCQ* reviewers for their advice at various stages in this article's production.

REFERENCES

Abramovitz, M., & David, P. A. (2000). American macroeconomic growth in the era of knowledge-based progress: The long-run perspective. In S. Engerman & R. E. Gallman (Eds.), *The Cambridge economic history of the United States: Vol. 3. The twentieth century* (pp. 1–92). Cambridge, England: Cambridge University Press.

Alcaly, R. (2003). *The new economy and what it means for America's future.* New York: Farrar, Straus & Giroux.

Baran, P. A., & Sweezy, P. M. (1966). *Monopoly capital: An essay on the American economic and social order.* New York: Monthly Review Press.

Bjkernes, G., & Bratteteig, T. (1995). User participation and democracy: A discussion of Scandinavian research on system development. *Scandinavian Journal of Information Systems, 7,* 73–97.

Braverman, H. (1988). *Labor and monopoly capital: The degradation of work in the twentieth century.* New York: Monthly Review Press. (Original work published 1974)

Brenner, R. (2002). *The boom and the bubble: The U. S. in the world economy.* New York: Verso.

Coogan, D. (2002). Public rhetoric and public safety at the Chicago Transit Authority: Three approaches to accident analysis. *Journal of Business and Technical Communication, 16,* 277–305.

Gould, J. R. (1964). *Opportunities in technical writing.* New York: Vocational Guidance Manuals.

Grossberg, L. (1992). *We gotta get out of this place: Popular conservatism and postmodern culture.* New York: Routledge & Kegan Paul.

Hall, S. (1988). *The hard road to renewal: Thatcherism and the crisis of the left.* New York: Verso.

Hardt, M., & Negri, A. (2000). *Empire.* Cambridge, MA: Harvard University Press.

Hardt, M., & Negri, A. (2004). *Multitude: War and democracy in the age of empire.* New York: Penguin.

Harvey, D. (1990). *The condition of postmodernity: An inquiry into the origins of cultural change.* New York: Blackwell.

Henry, J. (2001). Writing workplace cultures. *College Composition and Communication, 52.* Retrieved December 7, 2004, from http://archive.ncte.org/ccc/2/53.2/henry/article.html.

Henwood, D. (2003). *After the new economy.* New York: New Press.

Herndl, C. G. (1993). Teaching discourse and reproducing culture: A critique of research and pedagogy in professional and non-academic writing. *College Composition and Communication, 44,* 349–363.

Herndl, C. G. (1996). Tactics and the quotidian: Resistance and professional discourse. *Journal of Advanced Composition, 16,* 455–470.

Herndl, C. G., Fennell, B. A., & Miller, C. (1991). Understanding failure in organizational discourse: The accident at Three Mile Island and the Shuttle Challenger disaster. In C. Bazerman & J. Paradis

(Eds.), *Textual dynamics in the profession: Historical and contemporary studies of writing in professional communities* (pp. 279–305). Madison: University of Wisconsin Press.

Jameson, F. (1991). *Postmodernism; or the cultural logic of late capitalism.* Durham, NC: Duke University Press.

Johnson, R. R. (1998). *User-centered technology: A rhetorical theory for computers and other mundane artifacts.* Albany: State University of New York Press.

Longo, B. (2000). *Spurious coin: A history of science, management, and technical writing.* Albany: State University of New York Press.

Mandel, E. (1978). *Late capitalism* (J. De Bres, Trans.). New York: Verso. (Original work published 1972)

Marx, K. (1976) *Capital: Vol. 1.* (B. Fowkes, Trans.). New York: Penguin. (Original work published 1867)

Pine, J., II, & Gilmore, J. H. (1999). *The experience economy: Work is theatre and every business is a stage.* Boston: Harvard Business School.

Postrel, V. (2003). *The substance of style: How the rise of aesthetic value is remaking commerce, culture, and consciousness.* New York: HarperCollins.

Redish, J. (1995). Adding value as a professional communicator. *Technical Communication, 42,* 26–39.

Sauer, B. (2003). *The rhetoric of risk: Technical documentation in hazardous environments.* Mahwah, NJ: Erlbaum.

Schriver, K. (1993). Quality in document design: Issues and controversies. *Technical Communication, 40,* 241–257.

Schriver, K. (1997). *Dynamics in document design: Creating texts for readers.* New York: Wiley.

Schumpeter, J. (1942). *Capitalism, socialism, and democracy.* New York: Harper Perennial.

Schumpeter, J. (1982). *Business cycles: A theoretical, historical, and statistical analysis of the capitalist process* (Vol. 1). Philadelphia: Porcupine Press. (Original work published 1939)

Scott, J. B. (2003). Extending rhetorical-cultural analysis: Transformations of home HIV testing. *College English, 65,* 349–367.

Spinuzzi, C. (2002). A Scandinavian challenge, a U.S. response: Methodological assumptions in Scandinavian and U.S. prototyping approaches. In *Association for Computing Machinery, Special Interest Group Documentation 2002* (pp. 208–215). Toronto, Ontario, Canada: Association for Computing Machinery.

Street, J. (1992). *Politics and technology.* New York: Guilford.

Sullivan, D. (1990). Political-ethical implications of defining technical communication as a practice. *Journal of Advanced Composition, 10,* 375–86.

Thompson, E. P. (1966). *The making of the English working class.* New York: Vintage.

Thompson, J., DeTienne, K. B., & Smart, K. L. (1995). Privacy, email, and information policy: Where ethics meets reality. *IEEE Transactions on Professional Communication, 38,* 158–164.

Waddell, C. (1995). Defining sustainable development: A case study in environmental communication. *Technical Communication Quarterly, 4,* 201–216.

Williams, R. (1977). *Marxism and literature.* Oxford, England: Oxford University Press.

Yates, J. (1989). *Control through communication: The rise of system in American management.* Baltimore: Johns Hopkins University Press.

Mark Garrett Longaker is an assistant professor in the Division of Rhetoric and Writing at the University of Texas at Austin. He often contemplates local incomplete construction projects begun by now defunct tech firms.

Culture and Cultural Identity in Intercultural Technical Communication

R. Peter Hunsinger
Iowa State University

Drawing from the critical cultural theory of Arjun Appadurai, this article interrogates the concept of culture underpinning much intercultural technical communication research. Appadurai suggested that intertextual connections between the cultural and the economic, political, demographic, and historical aspects of the globalizing world are essential for understanding cross-cultural communication. The cultural theory offered in this article opens the way for further cultural studies research to be of use in intercultural technical communication theory, research, and pedagogy.

> The search for certainties is regularly frustrated by the fluidities of transnational communication. … Culture becomes less what Pierre Bourdieu would have called a habitus (a tacit realm of reproducible practices and dispositions) and more an arena for conscious choice, justification, and representation.
>
> Arjun Appadurai (1996), *Modernity at Large*, p. 44

Although a well-documented "explosion of interest in international professional communication" (Lovitt, 1999, p. 1) has inspired a body of creative scholarship in intercultural technical communication, the theoretical concepts of culture or cultural identity have yet to be examined critically for their impact on our disciplinary exchanges and activities. More important, perhaps, the notorious difficulties inherent in discussing the contested field of culture, such as oversimplification, essentialism, or ethnocentrism (see Goby, 1999; E. Weiss, 1998), remain a perennial problem for intercultural research. Such persistent problems and the limited theoretical reflection on the concept of culture, I believe, are interconnected issues: Difficulties in studying culture, I argue, stem from a problematic theoretical framework based largely in cultural heuristics and ethnographic descriptions that place too high a value on locating definitive culture. Working in the anthropological and sociological thread of cultural studies, however, reseachers can develop a more critical, flexible way of looking at culture and the cultural.

In this article, I draw from the influential work of anthropologist and sociologist Arjun Appadurai to interrogate what I take to be the predominant approach to researching and teaching intercultural technical communication, using a term I borrow from Beamer (2000), the *heuristic* approach (p. 113). After defining the contours of this approach, I describe how its implicit theory of culture and cultural identity structurally encourages teachers and researchers to overlook crucial aspects of cross-cultural communication. I then offer Appadurai's cultural theory as an alternative that allows for more critical intercultural research and pedagogy. Finally, I sketch broadly the implications that Appadurai's insights might have for the theory, research, and pedagogy of intercultural technical communication.

THE HEURISTIC APPROACH
TO INTERCULTURAL COMMUNICATION

At the outset, this article assumes that cultural identity is an important area of study for cross-cultural research and pedagogy. It therefore does not support a culture-free approach to intercultural communication, one that tries to identify "certain communication skills [that] are needed in all cultural settings" (Goby, 1999, p. 181). Although the culture-free approach may, in fact, skirt the problem of culture altogether, it raises its own set of difficulties. Edmond Weiss (1998) argued that universal expectations for various features of communication, such as clarity and persuasiveness, are hard to come by under any circumstances (p. 255); most attempts at universality have ended in ethnocentrism, no matter the intentions behind them. Importantly, too, the culture-free approach provides researchers and students little opportunity to explore the important roles of culture and cultural identity in communication.

If researchers accept that cultural identity and cultural differences do, in fact, play a significant part in cross-cultural communication, their object of inquiry should be the heuristic approach, which collectively describes the basis of most intercultural research and pedagogy. Working from catalogues of ethnographic data, this approach identifies important dimensions of culture and then rates particular cultures for each dimension, with the goal of developing workable descriptions that practitioners might find helpful in cross-cultural communication. For example, cultures can be classified as individualist or collectivist (a point of difference between North American and East Asian cultures, respectively), or masculine or feminine, and so on (Beamer, 2000). Decisions about the shape of cross-cultural communication are then, ideally, based on these cultural representations.

Beamer (2000, pp. 111–112) traced the heuristic approach to Hall's (1976) work in the 1970s and Hofstede's (1984) work in the 1980s. By the early 1990s,

Hall especially was being cited to help develop pedagogy specifically for intercultural technical communication (e.g., see Bosley, 1993; Thrush, 1993). By the end of the 1990s, as Edmond Weiss (1998) pointed out, "almost every [technical communication] course include[d] a unit on Edward Hall's high-context/low-context communication model ... [and] Hofstede's dimensions of culture seem[ed] to have become as commonplace" (p. 262). Even without explicit acknowledgement of the heuristic models of Hall, Hofstede, or the others that have appeared since then (e.g., Trompenaars & Turner, 1997), intercultural technical communication generally emphasized the heuristic approach to discuss the dimensions of culture (see Constantinides, St. Amant, & Kampf, 2001; DeVoss, Jasken, & Hayden, 2002). Now, midway through the following decade, the heuristic approach continues to pervade much of the intercultural technical communication research.

Taking their lead from the research, most technical communication textbooks that tackle intercultural communication have used the heuristic approach as well. Some texts, such as Markel's (2001) *Technical Communication* or Andrews's (2001) *Technical Communication in the Global Community*, focused on Hall's (1976) popular high-context/low-context dimension of culture. Burnett's (2005) *Technical Communication* offered a more thorough listing of cultural dimensions in a full-page chart that lists the continuum of cultural characteristics for each dimension (p. 54). Other texts, such as Anderson's (2003) or Pfeiffer's (2002), did not refer to Hall or Hofstede's (1984) cultural dimensions but provided a similar list of cultural variables for the technical communicator to note. All of these textbooks shared the assumption that describing particular cultures in terms of these heuristics will help the practitioner communicate effectively across a wide range of cultures.

SHORTCOMINGS OF THE HEURISTIC APPROACH

Though the heuristic approach has pervaded much intercultural communication research and pedagogy, its limitations have been well documented. Crystallizing a common complaint, Edmond Weiss (1998) noted that the heuristic approach "treats members of a group as instances of a profile," an essentializing practice that displaces cultural identity from the concrete individual into a typical instance of the individuals who share a culture (p. 260). Beyond simply typifying cultural identity, the heuristic approach is also prone to misrepresent cultural identity to emphasize what Munshee and McKie (2001) called the "differences that matter," and flattens culture to the reduced dimensions of the heuristic (p. 16). Reflecting a similar criticism, Beamer (2000) noted that the heuristic approach, if based on limited research or unrepresentative ethnographic data, can be misleading, though her purpose in pointing this out was to begin to synthesize a more precise heuristic.

Although I agree with these criticisms, I do not believe the way to avoid the problems with the heuristic approach is to develop a better researched, more detailed heuristic, as Beamer (2000) suggested. The problems in the heuristic approach run deeper than insufficient research or underdeveloped heuristics; underlying the heuristic approach is a problematic theory of culture and cultural identity. Culture is commonly treated as a prediscursive, effectively autonomous essence posing as a set of durable habits and practices, and cultural identity is something brought to communication rather than constructed and mobilized during communication. Culture and cultural identity, in other words, are allowed little flexibility and dynamism. For example, much of the intercultural communication research that attempts to define culture describes it as "an established set of values and a way of thinking that is passed from generation to generation" (Bosley, 1993, p. 53; for similar examples from textbooks, see Burnett, 2005, p. 41; Lay, Wahlstrom, Rude, Selfe, & Seltzer, 2000, p. 27; Markel, 2001, p. 127), and cultural identity is, similarly, relatively stable; other authors have offered variations on this definition of culture (see, e.g., Hein, 1991; Thrush, 1993; Warren, 2002).

This theory of culture and cultural identity did not develop specifically within intercultural technical communication, but rather comes from some of the primary sources of the heuristic approach: Hall and Hofstede. Although the purpose of studying culture for Hall (1976) was to "gradually free oneself from the grip of unconscious culture" (p. 211), he noted that culture is essentially an "irrational force" (p. 187) and that "deep cultural undercurrents structure life in subtle but highly consistent ways that are not consciously formulated" (p. 9). Similarly, Hofstede's (1984) extensive empirical study defined culture as "the interactive aggregate of common characteristics that influence a human group's response to its environment. Culture determines the identity of a human group in the same way as personality determines the identity of an individual" (p. 21). Both Hall's and Hofstede's definitions imply a theory of culture and cultural identity in which these two things are effectively stable (note the words *structure, consistent, determines*, and *identity*).

The overarching theoretical problem with which to begin is the insufficient separation between "culture" and "cultural." Note, for example, Hofstede's (1984) aforementioned claim that "culture determines the identity of a human group." Here, culture and cultural identity are inextricably linked in a causal relationship, where the latter is merely a manifestation of the former. In intercultural technical communication research specifically, the heuristic approach makes little distinction between culture and cultural, attempting to describe cultural identity by defining culture. The difference between the two terms, however, is crucial. As Appadurai (1996) explained, "If *culture* as a noun seems to carry associations with some sort of substance ... *cultural* the adjective moves one into the realm of differences, contrasts, and comparisons" (p. 12). In other words, when researchers fail to separate culture and the cultural, the noun *cul-*

ture implies a thing that can be positively located and described beneath the behaviors of certain identified groups of people, no matter how many hedges a researcher builds up around a cultural description. The result, then, is an approach
that describes individuals in terms of the typified cultural profile that Edmond
Weiss (1998) referred to in the quote cited earlier. For example, Andrews' (2001)
textbook cited an expert who linked the cultural behaviors observed in Japanese,
German, and French organizations to respective national cultures, which were
then plotted on a "high-trust/low-trust" continuum (pp. 9–10).

Dragga's (1999) "Ethical Intercultural Communication: Looking Through the
Lens of Confucian Ethics," though not referring to the heuristic approach, more
clearly illustrated this lack of separation between culture and the cultural. Dragga
described the Chinese system of Confucian ethics that influences Chinese cultural
characteristics, briefly outlining Confucius' central tenets, such as righteousness,
goodness, reverence, and so on. Dragga also discussed some challenges to Confucian ethics, from the ancient philosophy of Lao Tzu to Maoist thought and some of
the more contemporary Western influences. But although he was careful to explain
the interactions of Chinese cultural traditions with Western ones, Dragga still restricted Chinese culture to an ancient, specifically Chinese philosophy; influences
on culture after 1800, including the 1949 Communist Revolution, were reserved
for two paragraphs toward the end of the description (p. 374). I should be clear that
Dragga's discussion of Chinese culture was not misleading, ethnocentric, or inac-
curate. In fact, it presented a rather flexible and unique perspective on Chinese culture. My point, rather, is to show that Dragga generally anchored Chinese cultural
behavior to a system of ethics specific to the Chinese culture, and thus implied that
cultural identity must be tied to a substantive culture.

To their credit, some scholars who rely on the heuristic approach have gone to
great lengths to note that culture and cultural identity are not clear-cut or essential,
but fuzzy-edged and incomplete. Such scholars have often argued against
essentializing or oversimplifying culture or cultural identity when discussing the
heuristic approach. Burnett (2005), for example, noted that "binaries tend to treat
cultures as monolithic, that is, assuming that all Brazilian citizens are alike, ...
which can lead to stereotyping extremely diverse national and organizational cultures" (p. 53). The rough edges of cultural identity are made clear in caveats of this
type. However, although these hedges are certainly laudable and necessary for providing a fair portrait of culture and cultural identity, they function, in fact, as normative anti-essentialism; admitting grey areas in cultural description mostly serves
to assimilate anomalies into a profile, as hedges often justify making rhetorically
safe generalizations about culture and cultural identity. For example, Burnett followed the quote above by saying that the heuristic dimensions serve as a "useful
starting place" for studying culture (p. 53).

The discursive effects of the intercultural research, moreover, speak differently
than the hedges: Culture still acts as a thing, at least within the intercultural com-

munication situation. Even if the researcher acknowledges that cultural identity is tentative, fluid, and nonessential, the effects of tracing cultural identity to a culture are no less real, though the researcher may not be explicitly concerned about the ontological existence of a culture itself. Let me outline at least two ways in which, I believe, the lack of separation between culture and the cultural encourage an essentialist, effectively autonomous understanding of culture.

Lack of Intertextual Connections That Help Shape Cultural Identity

Because cultural identity tends to be traced to culture in the heuristic approach, the intertextual connections between cultural identity and other factors (e.g., economic or political) are often neglected; cultural identity is instead delimited in a prediscursive, effectively autonomous culture. I should note that the term *intertextual* does not refer only to the inherent heterogeneity or hybridity of cultures, or culture–culture interaction, which Timothy Weiss (1993) already did a great service describing in *"The Gods Must Be Crazy*: The Challenge of the Intercultural." Rather, *intertextual* here should imply the connections between cultural identity and extracultural factors, interactions that have not been well documented. For example, Bosley's (2001) introduction to her collection of intercultural communication case studies identified "the rapid expansion of corporate interests worldwide" that makes the world a "global village" (p. 1). DeVoss et al. (2002) similarly justified their study of intercultural communication textbooks by noting that "with the increasing globalization of the marketplace, the United States is becoming more multicultural and active in international business than it has previously been" (p. 69). (Textbooks also note the importance of globalization and multiculturalism: Anderson, 2003, p. 6; Burnett, 2005, pp. 38–39; Lay et al., 2000, p. 4; or Markel, 2001, p. 107.) But although intercultural communication research and pedagogy recognizes economic globalization, the concept of cultural identity is not allowed an analogously globalized dimension; cultural identity is still assumed to be effectively autonomous and independent of the dynamism of globalization. In other words, despite the rapid flows of migration and international travel, media, and communication that mark a globalizing world, cultural identity is commonly understood to be generally self-contained, conspicuously independent of economic, political, or technological influences.

In a different context, Longo (1998) noted the consequences of partitioning cultural research from its contexts: "A view of culture that is limited within the walls of one organization does not allow researchers to question assumptions about technical writing practices because those practices are not placed in relationship to influences outside the organization under study" (p. 55). Analogously, researchers laboring under the heuristic approach may neglect the links between cultural identity and global contexts, the vectors of power, politics, history, and capital that elucidate the ways cultural identity functions in cross-cultural communication.

Neglect of the Construction/Mobilization of Cultural Identity During Communication

Second, if culture is not clearly distinguished from the cultural, the construction and mobilization of cultural identity during discursive exchange tends to be neglected. That is, if cultural identity is traced to an independent, prediscursive substance, or culture, the heuristic approach treats cultural identity as something revealed during communication rather than a process mobilized in light of extracultural or contextual aspects of communication. The extracultural may include contextual factors that have fallen outside the scope of cultural study, such as economic and political infrastructures that circulate capital and power unevenly among a variegated population. In their critique of the heuristic approach, Munshee and McKie (2001) warned that its limited view of cultural identity "ignores the social processes behind the construction of cultural differences" (p. 19), especially, I should add, the cultural differences constructed or emphasized almost spontaneously during communication. Such differences may include, for example, heightened cultural contrasts, such as those occurring when cultural heritage appears to be threatened by so-called foreign influences, or what amount to caricatures of local cultural performances, such as those commonly put on for the participants in the popular reality television show *The Amazing Race*.

Let me take an example of what I consider to be a representative disciplinary discussion of cultural identity from a well-referenced work in intercultural technical communication. Thrush (1993) noted that "by the year 2000, 29% of the [domestic] workforce would be made up of people who had moved here from other countries" (p. 272). This observation supported Thrush's call for more attention to intercultural technical communication instruction. However, her focus was almost exclusively on workplace culture; the influences that the economic, political, demographic, and social implications that immigration would surely have on the intercultural communication situation were absorbed under the category of cultural differences. In addition, returning for a moment to Dragga's (1999) article discussed earlier, one can note that Chinese cultural identity is not shown to be constructed, but rather seems to be inherited from history. Despite the clear impacts that Western-style market economics has had on Chinese cultural identity in the last three decades, Dragga traced Chinese-ness in part to ancient Confucian ethics. The ways people mobilize Chinese cultural identity and cultural differences in, say, something like nationalistic pride, were not considered.

But when cultural identity is considered to be integrally rooted in economic, political, and historical contexts, students and researchers in intercultural technical communication can study the broad intertextual factors that influence the shape of cultural identity. As the following section shows, understanding these factors will allow intercultural technical communication research and pedagogy the flexibility to account for the fluidities of the globalizing world.

Sanctioned Ignorance: The Impact of the Heuristic Approach

I have suggested that the theory of culture and cultural identity underlying what I group under the heuristic approach encourages cultural identity to be represented as effectively autonomous, independent of economic, political, and historical contexts. The resulting lack of attention to cultural intertextuality and the mobilization of cultural identity leads to a significant gap in the heuristic approach: The limited theory of culture and cultural identity produces in intercultural research and pedagogy what Spivak (1999) termed a "sanctioned ignorance" (p. x) of the globalizing world; that is, the treatment of cultural identity as intrinsically linked to a substantive culture structures intercultural research and pedagogy to neglect features of the globalizing world that significantly influence cultural identity in communication.

What emerges in the intercultural research and pedagogy is a myopic focus on culture that displaces economic, political, and historical factors that nevertheless affect cross-cultural communication. In place of these factors, the intercultural research and pedagogy installs *culture* as the term around which communication turns; the assumption seems to be that categorizing and cataloguing cultures and cultural differences will build an adequate model for studying cross-cultural communication. For instance, Markel (2001) subordinated political and economic contexts and relationships under "Cultural Variables" (p. 107).

TOWARD AN INTERTEXTUAL THEORY OF CULTURAL IDENTITY

The solution to these structural problems in the heuristic approach, however, should not discount cultural identity or assume that culture does not influence communication, but should look at culture and cultural identity differently. For such an alternative perspective, I turn to the work of Appadurai, whose critical anthropological work constituted a theoretical intervention into the study of culture. Appadurai (1996) proposed that cultural research focus on the cultural as an active, deterritorialized process rather than attend to *culture* as a noun referring to a set of passively acquired or inherited traits (p. 12). In this regard, Appadurai reversed the approach to studying culture found in cultural heuristics; rather than trying to identify the ways culture manifests itself in the cultural, Appadurai studied the ways the idea of culture is constructed from the apparently cultural. In other words, it is crucial to understand cultural identity, not as a tacit set of reproducible norms and conventions, but as what Appadurai (1996) called "an arena for conscious choice, justification, and representation" (p. 44) within the dynamics of the globalizing world. This theoretical basis for intercultural research and pedagogy emphasizes the importance of cultural identity in communication, combining such topics as cultural intertextuality and the mobilization of cultural identity, while bracketing any notion of substantive or essentialized culture. The economic, political, and historical elements absent from the heuristic approach become an integral aspect of Appadurai's theory of the cultural, and the issues wrapped up in intercultural communication are not displaced solely into the category of culture.

Understanding two important assumptions that underpin Appadurai's (1996, 2000) work will help us approach his theory of cultural identity. The first assumption is common to much of the work in cultural studies. As Johnson (1987) wrote, "Culture is neither an autonomous nor an externally determined field, but a site of social differences and struggles" (p. 39). Culture, that is, is always the site of something beyond simply culture itself. The other assumption, which underpin Appadurai's (1996, 2000) work specifically, is that imagination, in the sense of representation and image, is a legitimate and central form of social practice in the globalizing world. Two features of contemporary life have placed imagination at the forefront of much of human experience: advances in media technologies that rapidly disseminate images and symbols around the world; and migratory patterns that rapidly circulate populations, problematizing regional or ethnic determinants of cultural identity (the very fact of intercultural communication as a popular, growing field attests to the grand-scale intermixing of cultures). Appadurai (1996) argued that because increasingly intense media and migration create "a new order of instability in the production of modern subjectivities" (p. 4), imagination is "central to all forms of agency, is itself a social fact, and is a key component of the new global order" (p. 31). This is not to say that cultural identity is entirely unrecognizable or upended in globalization; rather, imagination has become a form of cultural work, a field of negotiation between local experience and dynamic global influences. Highlighting the emphasis that imagination places on representation and textuality in the study of culture, Appadurai (1996) called his take on cultural studies "the relationship between the word and the world" (p. 51).

From these basic assumptions, Appadurai's influential theory claims that even apparently autonomous or localized cultural identities (such as those generally described in the heuristic approach) are the result of the intersecting processes of globalization. In globalization, Appadurai (2000) noted "the apparent stabilities that we see," such as culture or cultural identity, "are, under close examination, usually our devices for handling objects characterized by motion" (p. 5). Cultural identity for Appadurai is thus a confluence of mobile and shifting streams of textuality—of political, ideological, economic, or ethnographic texts. "The new global cultural economy," he argued, "has to be seen as a complex, overlapping, disjunctive order" (Appadurai, 1996, p. 32), characterized not by the interaction of relatively stable entities, such as effectively autonomous cultures, but rather by mobile worldwide currents that move independently of one another to converge and interact in a complex global system.

Appadurai (1996) characterized five global flows of textuality that make cultural identity irreducibly intertextual and unstable: *ethnoscapes*, *mediascapes*, *technoscapes*, *finanscapes*, and *ideoscapes*, which roughly correspond to movements of people, media, technologies, capital, and political ideologies, respectively. Sociologist Malcolm Waters (1995) later added the term *sacriscapes* to describe global flows of sacred values, which, with the rise of various forms of fundamentalism throughout the world, have grown to be important to much cultural interaction. Appadurai (1996) noted that the suffix *-scape* attached to these six terms demonstrates their "fluid and irregular shapes" (p. 33) and deeply perspectival nature. Individuals and groups in the globalizing world then draw from and respond to these scapes to constantly construct, reconstruct, and mobilize their cultural identities. It is important to note that the ways that individuals experience the interactions and forms of these currents in everyday life determines what Appadurai (1996) termed the "imagined worlds" (p. 33) in which people experience stability amidst ceaseless movement. The term *imagined worlds* evokes the importance of image and representation in shaping cultural identity, as well as the fact that cultural identity must always be "historically situated" among global cultural flows (Appadurai, 1996, p. 33).

An example based on the experience of a friend of mine, Lynn, who is from mainland China, can more concretely illustrate the ways these scapes build people's imagined worlds. Lynn came from China to a large Midwestern public university by tracing finanscapes and ideoscapes to a specifically American image of prestige and prosperity, and then followed a flow of immigration to the United States. She was well connected to the Chinese student community on campus, though she also kept a few close American friends; she thus often found herself shifting between different ethnoscapes. Lynn communicated with family and friends in China by using calling cards, Internet calling, e-mail, and a Webcam, taking advantage of the technoscapes and mediascapes that provided her contact with her home. Mediascapes, in fact, constituted an important part of her imagined world, as she drew many of her beliefs and expectations about what she understood to be American culture from Chinese-subtitled American movies she had seen in China. For instance, she was sometimes nervous to meet friends in local bars, because bars tended to be scenes of violence and crime in the American movies she had seen. In the United States, her relationship with American cultural production expanded into television, and she grew to appreciate sincerely the crassness of daytime talk shows, though she also watched Korean, Taiwanese, and Chinese movies and television shows she located on Chinese Internet message boards. To illustrate the complexity of these collisions of mediascapes and ethnoscapes, I can recall one instance watching a Korean movie with her as she read the Chinese subtitles, translating them into English for me.

Lynn's imagined world was thus the site of a complex intersection of textual scapes, from which Lynn constructed the cultural identity of a young Chinese woman in the United States. In other words, she did not simply express a cultural identity that she drew from an inherited, relatively stable Chinese culture, but rather constructed and mobilized a cultural identity performatively. For instance, Chinese-ness for her was at times self-conscious, as when she gave her close American friends traditional Chinese gifts (e.g., a fan, a double-happiness charm) during her first American Christmas celebration; she later explained to me that she was performing Chinese-ness in an American context, as she had never celebrated Christmas before and the Chinese rarely have a need to give each other relatively common Chinese objects. More generally and somewhat ironically, she felt that her Chinese cultural identity was stronger among her American friends, that is, when her perspective of the ethnoscapes, ideoscapes, mediascapes, and sacro-scapes shifted significantly among the different groups of people. Although her identification with her Chinese cultural identity remained essentially stable across different situations, her perception of the scapes deeply influenced the shape her cultural identity took in her interactions.

The intertextual construction and mobilization of cultural identity, however, is not limited to immigrant populations or a set of cosmopolitan individuals, but is a basic feature of the production of cultural identity. That is, as Appadurai (1996) noted, even in apparently local, stable, and autonomous cultural situations "locality must be maintained carefully against various kinds of odds," for example, through the discursive construction of barriers, identifications, and exclusions that divide inside from outside (p. 179). As Williams (1977) noted, the modern concept of culture itself developed to preserve a sense of stability amidst the chaotic up-heavals of the 19th and 20th centuries, and globalization has only exacerbated these social instabilities. Cultural identity thus becomes a process of mobilization that depends to a large extent on extracultural factors rather than on a kind of inher-ent cultural essence. When cultural identity is considered to be constructed, mobi-lized, and irreducibly intertextual, the intertextual connections that influence cul-tural identity during communication become significantly more important for understanding cross-cultural communication.

IMPLICATIONS: INTERCULTURAL THEORY, RESEARCH, AND PEDAGOGY

Appadurai's (1996, 2000) insights into the intertextual nature of cultural identity suggest a new constellation of problems to consider in intercultural technical communication, opening the field to more critical work in cultural studies to supplement the work begun under the heuristic approach. Let me sketch in broad, tentative strokes the ways that theory, research, and pedagogy might begin to study the ways cultural identity is intertextually constructed and mobilized during communication.

Implications for Theory and Research

Although every intercultural communication article does not need to present a fully developed economic and political critique, Appadurai's (1996, 2000) insights offer ways of discussing cultural identity to reflect the conflicts and complexities of the globalizing world. Let me point out two areas of intercultural technical communication research on which Appadurai's cultural theory can shed some light. First, the set of issues involved in cultural research can open to include factors that have fallen outside the study of culture in the heuristic approach; context can broaden to include the political, economic, and social issues surrounding shifting cross-cultural communication situations, expanding the limited focus on culture itself to global contexts.

For this expansion, Appadurai (1996) proposed that the very models of local or autonomous culture should change, because globalization has made it difficult to characterize cultures definitively in any useful way. Instead, researchers should not expect cultures to exhibit Euclidean boundaries or structures, but rather address the specifically dynamic problems that stem from cultural conflict or confusion (Appadurai, 1996, pp. 46–47), attending to the cultural rather than to culture. Instead of illustrating, for example, how North American and Japanese notions of power–distance present problems in business relationships, researchers might ask why cultural differences are a problem in the first place: Is the problem simply that the two cultures are incompatible in terms of power–distance, or do other issues—political, economic, or demographic—in some way make cultural differences a site of unnecessary difficulty? The study of cultural conflicts such as this would do well to look to factors other than culture itself to characterize the communicational difficulty adequately. For this agenda, intercultural theory and research might follow the lead of cultural studies, which, Nelson, Treichler, and Grossberg (1992) pointed out, has a long history of interdisciplinarity and more nuanced treatments of culture.

A focus on the cultural also suggests that researchers study the ways cultural issues reflect many of the underlying antagonisms of the globalizing world, because what appear to be cultural issues often extend into political or social issues. For example, Zizek (2000) pointed out that "in today's political discourse, the term 'worker' has disappeared, supplanted or obliterated by 'immigrants' ... , [which transforms] the class problematic of workers' exploitation ... [into] the multiculturalist problematic of the 'intolerance of Otherness' " (p. 10). Often, that is, problems stemming from several factors tend to be articulated in terms of conflicts that are narrowly defined as cultural. The role of the intercultural researcher, in this case, would be to explore the ways underlying social contexts exacerbate or even shape the cultural conflict, considering issues that may lie outside culture itself.

Second, if cultural stability is something produced and reinforced, as Appadurai (1996, 2000) argued, the performative role of the intercultural researcher in producing a stable, definable culture warrants self-reflexive examination. Noting the precarious place of the cultural researcher, Appadurai (1996) wrote

> The ethnographic project is in a peculiar way isomorphic with the very knowledge it seeks to discover and document, as both the ethnographic project and the social project it seeks to describe have the production of locality as their governing telos. (p. 182)

Similarly, Bourdieu (2001) explained that cultural description is essentially performative because describing the social world "aims to produce and impose representations (mental, verbal, visual, or theatrical) of the social world which may be capable of acting on this world by acting on agents' representation of it" (p. 127). Thus, by creating representations that, if pedagogically useful, are to be enacted, cultural researchers produce the cultural locality and autonomy they intend to describe, producing the object of study in the act of research. A researcher undertaking a self-reflexive inquiry might begin with Herndl's (1991) "Writing Ethnography," which explores the rhetoricity of qualitative ethnographic research, and which could readily be applied to the intercultural research problematic. Henry's (2000) ethnographic study can also serve as a precedent for the kind of self-critical scholarship I suggest, particularly for those who would study the ways academic textual practices perpetuate certain modes of representing intercultural exchanges in the workplace.

Implications for Pedagogy

In addition to opening new areas of intercultural technical communication research, Appadurai's (1996) cultural theory can show technical communication students the intertextual influences that shape the construction and mobilization of cultural identity during communication. Ultimately, students exposed to this kind of pedagogy would better understand the nature of communication in cross-cultural contexts against the background of globalization. Studying the intertextual nature of cultural identity, however, should not simply be considered a politically correct (i.e., anti-essentialist) but impractical exercise. Rather, a focus on extracultural factors in a course or unit on intercultural technical communication can be a productive and practical way to approach cultural issues in the classroom. Aronowitz (2000)—whose popular book *The Knowledge Factory* rails against the job-training curriculum that he said pervades most universities—supported an expanded notion of the practical; instrumental pedagogies designed for skills training, he argued, "have failed to prepare students to face relatively new issues such as globalization, immigration, and cultural conflict" (p. 127). "Ironically," he noted, "the best preparation for the work of the future might be to cultivate knowledge of the broadest possible kind, to make learning a way of life that in the first place is pleasurable and then rigorously critical" (p. 161).

Aronowitz's (2000) path to this kind of broad, critical learning is to help students see the necessary conflicts in the knowledge production process. In an intercultural pedagogy designed to interrogate Western biases in materials for an intercultural-communication course, for example, Munshee and McKie (2001) had students read selections from critical cultural theorists and postcolonial scholars who offered a perspective on intercultural exchange different from that often found in professional communication materials. Students might read, for example, selections from Said's (1978) *Orientalism* to learn how Western scholars have studied other cultures to better administer the European colonies; students could be encouraged to explore parallels between imperialistic Orientalist scholarship and intercultural communication research, exposing the economic, political, and historical factors underlying intercultural exchange. Films such as Volkmer's (1994) *The Bomb Under the World* can also illustrate in sharp relief alternative perspectives on globalization and the cultural, and may prove more accessible to students. In this film, an Indian village that views itself as removed from the dynamics of the globalizing world finds itself at the crossroads of ethnoscapes, finanscapes, and technoscapes, and influenced by a specifically Western way of living. The point of this pedagogy is to examine the world in its various inequalities, imbalances, and asymmetrical economic and political relationships, to supplement what the heuristic approach tends to exclude from the study of cultural identity in communication. In terms of Appadurai's (1996) cultural theory, these critical readings supply pictures of the ethnoscapes, technoscapes, mediascapes, ideoscapes, finanscapes, and sacriscapes of the globalizing world that instructors and students might draw from to develop an intertextual model of intercultural communication.

As is becoming apparent, accepting the implications of Appadurai's (1996) cultural theory means that instructors will need to supplement the current textbooks that cover cross-cultural issues. This is not to say that textbooks, which have been a rich if sometimes unfair target of much criticism (Miles, 1997), have ultimately failed to address the cultural. Rather, the textbooks' perspective on intercultural communication should be one such view offered in the classroom to teach an intertextual approach to cultural identity. In the pedagogy I suggest, to borrow from the documentary *bell hooks* (Jhally, 1997), "the issue is not freeing ourselves from representations," even cultural ones that seem skewed or incomplete. hooks continues, "It's really about being enlightened witnesses when we watch representations." To illustrate how current course materials might work with critical supplementary texts, let me take an example from Markel's (2001) *Technical Communication*. In a section titled "Ethics and Multicultural Communication," Markel offered a case study that models the ethical dilemmas technical writers might confront when interacting with individuals of different cultures (pp. 39–40). McNeil at Informatics is considering submitting a proposal to Crescent Petroleum, a Saudi Arabian oil-refining corporation, to develop a company intranet. At a briefing in New York, Denise McNeil is struck by what she sees as sexist Saudi business practices: The Crescent representatives seem "uncomfortable" with McNeil and do not socialize with her during a break. Identifying the Saudi company as "traditional," McNeil decides to submit a proposal anyway but conceal that she had founded the company and that she and another woman hold the positions of president and chief financial officer. Making a cultural inference, she also considers hiding the lead engineer's name, "Feldman." Students are then asked how they might respond to this situation ethically.

A more important issue to me, however, is the way the case study represents Saudi cultural practices. The structure of the dilemma is an us-versus-them scenario in which the student's role is to deal with odd, foreign, and (from a liberal Western perspective) negative cultural customs. This is not to say that the Saudis are necessarily demonized or depicted inaccurately. However, the description of traditional Saudi cultural practices traces these characteristics to an inaccessible tradition beyond critical examination, or, in other words, to a substantive and inherited culture. The intertextual connections that have shaped Saudi cultural practices, which could dissolve the us-versus-them opposition, are not considered. In this case, an article from the *New York Times Magazine* can supplement the course material to draw out some of these intertextual connections. In "The Jihadi Who Kept Asking Why," Rubin (2004) claimed that the treatment of women under Islamic law, which roughly correlates to the Saudis' sexist treatment of McNeil in Markel's (2001) case study, is a reactionary cultural practice, a feature of the radically conservative revolution brought on by Wahhabi clerics in response to the Westernizing oil boom of the 1970s (pp. 41–3). In Appadurai's (1996) terms, ideoscapes and sacriscapes (fundamentalist politics) have converged with finanscapes and mediascapes (oil money and Western culture) to shape Saudi cultural practices into the ones depicted in the case study. The traditional, in large part, has been invented against the context of these extracultural factors.

Moreover, understanding the intertextual connections between cultural practices and extracultural factors can shed some light on how Saudi cultural identity might be mobilized during communication. For instance, the popular opposition to the second Iraq war across most of the Arab world, including Saudi Arabia, may encourage the case study's Saudi businessmen to accentuate their traditional cultural identity in a minor act of resistance or even in solidarity with increasingly influential fundamentalist clerics; sacriscapes and ideoscapes may thus intersect in the briefing in New York. Finanscapes might also come into play, as the heavy U.S. reliance on Middle Eastern oil and investment can allow the Saudis to be more brazen with cultural practices that Westerners tend to find offensive. At any rate, the case study alone, which focuses on the cultural "differences that matter" and traces them back to an effectively autonomous, supposedly traditional culture, does not provide the opportunity for critical inquiry into the cultural differences it describes, leaving students with a static cultural representation. To interrogate the ways cultural practices are intertextually constructed and mobilized for certain purposes, then, the textbook description must be supplemented, as I have begun to model here.

CONCLUDING NOTES

Even if cultural issues do not revolve around the noun *culture* as the master signifier that describes every difficulty arising in cross-cultural communication, the cultural will always be immensely difficult to research and teach. However, the cultural is too important to leave to chance or intuition; culture is one of those volatile uncertainties that ironically holds sway over so much human interaction, especially as cultural differences have become a battleground for the economic and political issues in the globalizing world. The cultural, then, must continue to be interrogated critically so that those working in intercultural technical communication might surpass the more devastating angels of our nature and interact flexibly and effectively on the global scene. This article attempts to consider new ways of approaching cultural issues in technical communication, to equip researchers and teachers with a place from which to study culture. The theoretical framework offered here must be enriched, developed, and specified before it can begin to answer many of the profound questions we as a discipline have about intercultural technical communication.

ACKNOWLEDGMENT

I want to thank the members of my master's thesis committee, Helen Rothschild Ewald, David R. Russell, and Mark Rectanus, for their invaluable guidance on the earliest drafts of this article, and especially Helen's critical eye on subsequent drafts. All faults are mine, but I attribute any of the article's saving graces to her generous advice.

REFERENCES

Anderson, P. V. (2003). *Technical communication: A reader-centered approach* (5th ed.). Canada: Thomson/Heinle.

Andrews, D. (2001). *Technical communication in the global community.* Upper Saddle River, NJ: Prentice Hall.

Appadurai, A. (1996). *Modernity at large: Cultural dimensions of globalization.* Minneapolis: University of Minnesota Press.

Appadurai, A. (2000). Grassroots globalization and the research imagination. *Public Culture, 12,* 1–19.

Aronowitz, S. (2000). *The knowledge factory: Dismantling the corporate university and creating true higher learning.* Boston: Beacon.

Beamer, L. (2000). Finding a way to teach cultural dimensions. *Business Communication Quarterly, 63,* 111–118.

Bosley, D. (1993). Cross-cultural collaboration: Whose culture is it, anyway? *Technical Communication Quarterly, 2,* 51–62.

Bosley, D. (Ed.). (2001). *Global contexts: Case studies in international technical communication.* Boston: Allyn & Bacon.

Bourdieu, P. (2001). *Language and symbolic power.* (G. Raymond & M. Adamson, Trans.). In J. B. Thompson (Ed.). Cambridge, MA: Harvard University Press.

Burnett, R. (2005) *Technical communication* (6th ed.). Boston: Thomson Wadsworth.

Constantinides, H., St. Amant, K., & Kampf, C. (2001). Organizational and intercultural communication: An annotated bibliography. *Technical Communication Quarterly, 10,* 31–58.

DeVoss, D., Jasken, J., & Hayden, D. (2002). Teaching intracultural and intercultural communication: A critique and suggested method. *Journal of Business and Technical Communication, 16,* 69–94.

Dragga, S. (1999). Ethical intercultural communication: Looking through the lens of Confucian ethics. *Technical Communication Quarterly, 8,* 365–381.

Goby, V. P. (1999). All business students need to know the same things! The non-culture-specific nature of communication needs. *Journal of Business and Technical Communication, 13,* 179–189.

Hall, E. T. (1976). *Beyond culture.* Garden City, NY: Anchor/Doubleday.

Hein, R. G. (1991). International technical communication. *Technical Communication, 38,* 125–127.

Henry, J. (2000). *Writing workplace cultures: An archaeology of professional writing.* Carbondale: Southern Illinois University Press.

Herndl, C. (1991). Writing ethnography: Representation, rhetoric, and institutional practices. *College English, 53,* 320–332.

Hofstede, G. (1984). *Culture's consequences: International differences in work-related values.* (Abridged ed.). Beverly Hills, CA: Sage.

Jhally, S. (Director). (1997). *bell hooks: Cultural criticism and transformation* [Motion picture]. (Available from Media Education Foundation, 60 Masonic Street, Northampton, MA 01060)

Johnson, R. (1987). What is cultural studies anyway? *Social Text, 16,* 38–80.

Lay, M. M., Wahlstrom, B. J., Rude, C. D., Selfe, C. L., & Selzer, J. (2000). *Technical communication* (2nd ed.). Boston: McGraw-Hill.

Longo, B. (1998). An approach for applying cultural study theory to technical writing research. *Technical Communication Quarterly, 7,* 53–73.

Lovitt, C. R. (1999). Introduction: Rethinking the role of culture in intercultural professional communication. In Lovitt, C. R. & Goswami, D. (Eds.), *Exploring the rhetoric of international professional communication: An agenda for teachers and researchers* (pp. 1–13). Amityville, NY: Baywood.

Markel, M. (2001). *Technical communication* (6th ed.). Boston: Bedford/St. Martin's.

Miles, L. (1997). Globalizing professional writing curricula: Positioning students and re-positioning textbooks. *Technical Communication Quarterly, 6,* 179–200.

Munshee, D., & McKie, D. (2001). Toward a new cartography of intercultural communication: Mapping bias, business, and diversity. *Business Communication Quarterly, 64,* 9–27.

Nelson, C., Treichler, P., & Grossberg, L. (1992). Cultural studies: An introduction. In C. Nelson, P. Treichler, & L. Grossberg (Eds.), *Cultural studies* (pp. 1–17). New York: Routledge & Kegan Paul.

Pfeiffer, W. S. (2002). *Technical writing: A practical approach* (5th ed.). Upper Saddle River, NJ: Prentice Hall.

Rubin, E. (2004, March 7). The jihadi who kept asking why. *New York Times Magazine*, 34–44, 62–64, 113–114.

Said, E. (1978). *Orientalism*. New York: Vintage.

Spivak, G. C. (1999). *A critique of postcolonial reason: Toward a history of the vanishing present.* Cambridge, MA: Harvard University Press.

Thrush, E. (1993). Bridging the gaps: Technical communication in an intercultural and multicultural society. *Technical Communication Quarterly, 2,* 271–283.

Trompenaars, F., & Turner, C. H. (1997). *Riding the waves of culture: Understanding diversity in global business* (2nd ed.). New York: McGraw-Hill.

Volkmer, W. (Director). (1994). *The bomb under the world* [Motion picture]. (Available from Bullfrog Films, P.O. Box 149, Oley, PA 19547)

Warren, T. L. (2002). Cultural influences on technical manuals. *Journal of Business and Technical Communication, 32,* 111–123.

Waters, M. (1995). *Globalization*. New York: Routledge & Kegan Paul.

Weiss, E. (1998). Technical communication across cultures: Five philosophical questions. *Journal of Business and Technical Communication, 12,* 253–269.

Weiss, T. (1993). The gods must be crazy: The challenge of the intercultural. *Journal of Business and Technical Communication, 7,* 196–217.

Williams, R. (1977). *Marxism and literature*. Oxford, England: Oxford University Press.

Zizek, S. (2000). *The fragile absolute, or, why is the Christian legacy worth fighting for?* New York: Verso.

R. Peter Hunsinger is a doctoral student at Iowa State University, where he researches cultural studies, globalization issues, and critical pedagogies. He received a master of arts degree from Iowa State University in 2004.

Disability Studies, Cultural Analysis, and the Critical Practice of Technical Communication Pedagogy

Jason Palmeri
Ohio State University

This article critically analyzes how technical communication practices both construct and are constructed by normalizing discourses, which can marginalize the experiences, knowledges, and material needs of people with disabilities. In particular, the article explores how disability studies theories can offer critical insights into research in two areas: safety communication and usability. In conclusion, the article offers ways that disability studies can intervene in the pedagogy of usability, communication technology, linguistic bias, narrative, and discourse communities.

In the past few years, cultural studies scholars (Henry, 2000; Herndl, 1993; Lay, 2000; Longo, 1998, 2000; Scott, 2003) have been exploring how technical communication practices both shape and are shaped by powerful social discourses. Drawing on Foucault, as well as other critical social theorists, these scholars have interrogated how technical communication participates in the discursive process of normalization: legitimating and subjugating knowledges, examining and controlling workplace practices, forming subjectivities, and marking bodies as normal or deviant. Arguing that the practices of technical communication often work to reinforce material social inequalities, cultural studies scholars have also demonstrated the need to intervene to contest and provide alternatives to technical communication's regime of normalization. In these calls for critical intervention, the emerging discipline of disability studies has rarely been cited as a potential source of theory; yet disability studies has much to offer, as it is centrally concerned with interrogating "the divisions our society makes in creating the normal versus the pathological" (Linton, 1998, p. 2).

Rejecting the conventional medical model of disability that focuses on rehabilitating individuals with disabilities so that they can fit into an ableist society,[1] dis-

---

[1]An ableist society is one that takes able-bodiedness as a norm and thus discriminates against people with disabilities. For more discussion of ableism, see Linton (1998).

ability studies theorists proffer a social/political model of disability that fore-grounds the need to adapt social discourses and material environments to ensure equal participation for citizens of diverse abilities. In interrogating the social con-struction of disability and normalcy, disability studies theorists ask questions that could usefully extend current critiques of the normalizing practices of technical communication.

1. How does the social construction of citizenship and subjectivity depend upon the othering of people with disabilities? (Garland-Thomson, 1997)
2. How does the historical practice of eugenics, which sought to define bodily norms and eradicate people with disabilities who were deemed deviant, continue to inform contemporary discourses of technology, science, medi-cine, and public policy? (Davis, 1995; Hubbard, 1997; Russell, 1998)
3. How do visual and verbal rhetorical practices work to reinforce and/or sub-vert social constructions of normalcy and disability? (Brueggemann, 1999; Garland-Thomson, 2002; Wilson & Lewiecki-Wilson, 2001)

Although disability studies theory has been relatively absent from the conversation about technical communication and normalization, technical communication scholars have contributed greatly to the literature on making texts accessible to in-dividuals with disabilities (Carter & Markel, 2001; Ray & Ray, 1998; O'Hara, 2004).[2] This line of research has focused primarily on educating technical commu-nicators about following or enhancing current standards for ensuring access.

Although these scholars have demonstrated how technical communicators may effectively address the concerns of users with disabilities, they have not explored ways in which technical communication discourse is enmeshed in the broader so-cial construction of disability and normalcy. Yet as technical communication in-creasingly constructs itself as a profession that assists users with disabilities, we researchers must begin to critically engage disability studies' critiques of how pro-fessional discourses often work to reinforce normalcy and marginalize the embod-ied knowledges of people with disabilities (Linton, 1998).

Wilson's (2000) "Making Disability Visible" has already begun the work of placing technical communication in dialogue with disability studies. In this article, Wilson demonstrated how a disability studies perspective can contribute to the pedagogy and theory of medical and scientific writing, arguing that "disability studies provides a unique site from which to critically examine the assumptions of medicine and science and their interrelated and mutually reinforcing discourses" (p. 151). Showing the relevance of disability studies for medical and scientific

---

[2]The Society for Technical Communicators' (STC) AccessAbility special interest group (http://www.stcsig.org/sn/index.shtml) has also published many highly valuable newsletters addressing issues of access for users and for members of the technical communication profession.

writing pedagogy, Wilson outlined a series of activities that encourage students to read critically the ways in which medicine and science socially construct disability and to incorporate the perspectives of people with disabilities into medical/science writing. Ultimately, I seek to extend Wilson's work by demonstrating how disability studies can also inform technical communication research and pedagogy outside of the medical and scientific realms.

Technical communication's general focus on workplace discourse practices makes it a vital area for exploring the material consequences of the social construction of disability. Indeed, all three titles of the Americans with Disabilities Act (ADA) address the access that people with disabilities have to employment and/or services in corporate, government, and nonprofit workplaces (O'Brien, 2004). Nevertheless, in an environment in which 93% of plaintiffs' ADA cases fail (Colker, 1999) and many key provisions of the law are being restricted by the courts, people with disabilities still face many barriers in accessing the workplace. In view of these material conditions, it is vital that technical communicators begin to

1. Interrogate critically the social discourses of disability (and normalcy) that work to constrain workplace access.
2. Consider ways in which technical communication practices both shape and are shaped by these discourses.
3. Imagine ways in which technical communicators, both those with and those without disabilities, can intervene in or transform these discourses.

Yet the need to increase the material access of people with disabilities is certainly not the only reason technical communicators should engage with disability studies perspectives. Technical communication and disability studies share many similar concerns and thus could productively inform one another. Both technical communication and disability studies advance a social constructionist view of science and technology—a view that emphasizes the need to interrogate the ethical, social, and political effects of scientific and technical discourse. In particular, disability studies shares with feminist technical communication a concern for critiquing how scientific and technical discourses participate in the social construction of bodies in ways that reinforce social hierarchies and marginalize certain kinds of knowledges.

To ground my discussion of disability studies in technical communication, I critically explore theory and research in two areas: safety communication and usability. I then conclude by offering several pedagogical interventions that can enable teachers to integrate disability studies throughout their courses. Before delving into this analysis, however, I would first like to position myself in relation to this topic. I identify as a temporarily able-bodied person—a positioning that recognizes that disability will likely be a part of my embodied experiences at some point in the future, and that in a few ways it already has in the past. I also would like to note that I have come to disability studies theory rather recently in my academic work. In an early study of collaborative, medically-oriented writing in a law firm, I completely elided concerns of the social construction of disability even though this issue was highly present in the texts I was studying. Furthermore, although I always addressed disability in perfunctory ways in course readings and activities and worked to provide accommodations for individual students, I have only very recently come to see how a disability studies lens could transform my entire pedagogical practice. Thus in critically rereading scholarship on safety communication and usability from a disability studies perspective, I am in some sense critically rereading my own work as well, seeking not so much to critique previous scholarship as to begin to open up the technical communication field to the numerous insights that I believe disability studies theory can provide.

THE EUGENIC UNCONSCIOUS OF SAFETY COMMUNICATION

Seeking to investigate areas of technical communication in which the regime of normalcy is reinforced, I critically explore how disability figures in safety communication discourse. By rereading Madaus' (1997) groundbreaking historical study of safety communication through a disability studies frame, I demonstrate ways in which the development of safety communication was ultimately enmeshed in normalizing/eugenic discourses that sought to marginalize or eradicate people with disabilities.[3]

Taking a feminist historiographic perspective, Madaus (1997) told the fascinating story of numerous women who founded the discipline of health and safety com-

---

[3]My argument here is indebted to Davis' (1995) work on the concomitant development of normalcy and eugenics in the 19th and 20th centuries. According to Davis, "the word 'normal' as 'constituting, conforming to, not deviating or differing from, the common type or standard, regular, usual' only enters the English language around 1840" (p. 24). Interestingly, the notion of normal was accompanied by the 19th century development of the statistical bell curve: "The norm pins down that majority of the population that falls under the arch of the standard-shaped bell curve .... When we think of bodies, in a society in which the concept of norm is operative, then people with disabilities will be thought of as deviants" (p. 29).

This concept of the norm (that could be tracked and measured) gave rise to the eugenics movement. Ultimately, the division of the bell curve "into quartiles, ranked order, and so on creates a new kind of 'ideal.' ... The new ideal of ranked order is powered by the imperative of the norm, and then supplemented by the notion of progress, human perfectibility, and the elimination of deviance, to create a dominating hegemonic vision of what the human body should be. Although people tend to associate eugenics with Nazi-like racial supremacy, it is important to recognize that eugenics was not the trade of a fringe group of right-wing fascist maniacs. Rather, it became the common practice of many, if not most, European and American citizens." (Davis, 1995, p. 35)

munication in the early part of the 20th century. In telling this story, Madaus sought to recover and value the voices of previously overlooked female technical communicators who critiqued the ideological implications of science and reimagined technical communication as a public, activist practice. Madaus aptly noted that these women were master rhetors who greatly modified their messages for scientific, governmental, and public audiences. In addition, the women combined appeals to scientific logos (statistics) with emotionally powerful narratives of workers who were harmed by industrial conditions. Drawing on the work of Waddell (1990) and Sauer (1993), Madaus singled out the use of emotion as a particularly important device for feminist rhetorical action: "The ability of emotional appeals to motivate policy change is often overlooked or devalued (Waddell, 1990). Eastman's work stands in mark contrast to contemporary accident reports which 'privilege the rational (male) objective voice and silence human suffering' (Sauer, 'Sense' 63)" (p. 265). Elucidating the powerful appeals in Eastman's (1910) *Work Accidents and the Law*, Madaus noted that Eastman persuasively demonstrated that over 509 men a year were seriously injured in Pittsburgh in industrial accidents and then "she increased the [emotional] impact of these figures by extrapolating a ten-year figure, 5,000, which prompted her to imagine a *little city of cripples*" (p. 264). Using the work of Garland-Thomson (2002) as a frame, one can move from seeing the "city of cripples" quote as an example of pathos to conceptualizing it as an example of the sentimental gaze that "produces the sympathetic victim or helpless sufferer needing protection or succor and invoking pity, inspiration, and frequent contributions …. In such appeals, impairment becomes the stigma of suffering, transforming disability into a project that morally enables a nondisabled rescuer" (p. 63).

This gaze is especially apparent in the photos taken by Lewis Hine that accompany Eastman's (1910) text, photos whose pathetic appeals Madaus (1997) praised. In one photo, a worker stands with one hand outside his overalls and one hand tucked within them; the caption reads, "THE WOUNDS OF WORK: When a man's hand is mutilated he keeps it out of sight" (Eastman, 1910, p. 144). In this way, the photo caption exalts able-bodiedness and invites readers to pity the subject for his unsuccessful attempt to achieve the appearance of normalcy. In another photo, a man with only one arm stands on a front porch with his family, all of whom appear very sorrowful. The caption reads, "One arm and four children" (p. 153). In this photo, the male worker is metonymically referred to by his disability—his one arm—and the audience is asked to pity him because he can no longer

Although the American eugenic movement did certainly have strong racist, sexist, and classist dimensions, it was also very centrally concerned with the identification and eradication of disability, through such methods as institutionalization and forced sterilization of women with disabilities. And although the eugenics movement is now officially discredited, its assumptions continue to inform many discursive and material practices that seek to eradicate disability (often through genetic means) rather than transform society to adapt to it (Hubbard, 1997; Russell, 1998).

provide for his family. In this way, Eastman's text advances the cause of industrial safety by evoking the disabled person as a sentimental object, putting all emphasis on preventing disability and overlooking the need to adapt social institutions to make the lives of those with disabilities more livable. In addition to evoking the sentimental gaze, Eastman also drew upon the eugenic gaze, which classifies people with disabilities as problems to be eradicated. Through extensive qualitative and quantitative analyses, Eastman outlined the great social costs of disability: "Every helpless cripple left an unwilling burden on those who can ill afford to support him is a burden upon society" (p. 166). By constructing a pathetic appeal that evoked horror at the great burden of a "city of cripples," Eastman ultimately gained power for her (admittedly important) cause of workplace safety by appealing to unconscious social desires to eliminate people with disabilities from the polis. Thus, although it is important to value Eastman's significant contributions to safety communication—as Madaus powerfully called for—scholars must also be careful to interrogate critically the eugenic unconscious of safety communication—the ways that safety and normalcy are inextricably bound.

This critical reevaluation of Eastman's (1910) historic use of disability images in safety communication is all the more pressing because disability continues to appear problematically in contemporary workplace safety reports. For example, Schlosser's (2002) highly persuasive exposé about the appalling labor conditions in meat-packing plants relates the case of Kenny, a worker who was severely injured numerous times. Schlosser wrote, "Once strong and powerfully built, he [Kenny] now walks with difficulty, tires easily, and feels useless, as though his life were over. He is forty-six years old" (p. 190). Effectively using the sentimental image of Kenny to dramatize the consequences of unregulated industry, Schlosser took for granted that Kenny would feel useless; he did not critically consider the social conditions that make it difficult for Kenny to imagine/live life with his disability.

As technical communicators increasingly take on the vital and important work of advocating for worker and consumer safety, we must be mindful that we do not unwittingly reinforce destructive images of the social implications of disability. To begin integrating disability studies critique into safety communication scholarship, we might ask the following questions:

1. How has safety communication historically relied on eugenic arguments about the need to mitigate the so-called economic burden of disability on society? To what extent does this eugenic unconscious persist in safety rhetoric today?

2. How do safety communication narratives draw upon problematic sentimental narratives of disability as an individual tragedy to be pitied? How could safety communication narratives integrate a social/political view of disability while still dramatizing the real harm that workplace accidents cause?

3. Where do people with disabilities appear (and not appear) in safety communication texts? Are people with disabilities relegated to the role of victims of poor safety procedures? How might safety communication texts address and be cowritten by people with disabilities as vital participants in the process of creating and maintaining safe environments and products?

4. What particular safety challenges do people with disabilities face as workers and consumers? How might attempts to make workplaces and products safer for people with disabilities result in insights that could improve safety more generally?

USABILITY DISCOURSE AND
THE TROUBLE WITH NORMALCY

Although issues of disability access have remained largely separate from the scholarly conversation in safety communication, usability scholars have been at the forefront of considering ways to increase access for users with disabilities. By conducting a close, disability studies-inflected reading of two foundational articles on usability and disability (Ray & Ray, 1998; Salvo, 2001), I hope to open up a conversation about how technical communication's rhetoric of disability accessibility may inadvertently reinforce problematic constructions of normalcy, even as it simultaneously increases access for people with disabilities in positive ways.

Technical Communication as Rehab Science

Ray and Ray (1998) began their discussion of designing for users with visual impairments by noting that "as technical communicators, we often assume that our audience will interact with information in the same way that we do: we see pictures, fonts, colors, links, and page layouts, and we read, interpret, or click them as appropriate. The fact is, though, that a significant portion of our audience—people with visual impairments—cannot readily interact with information in this way. Many visually impaired people require the help of adaptive technologies" (Ray & Ray, 1998, p. 573). Although Ray and Ray gave a very useful summary of the challenges often faced by visually impaired users in interacting with online texts (and also offer cogent suggestions for accessible design), their language implicitly constructs the *we* of technical communicators as people with *normal* vision, thereby subtly marginalizing blind and visually impaired people from the profession. In other words, this article implicitly conceives of technical communication as a kind of rehabilitative profession of experts who assist those with disabilities in accessing information. In this way, disability is constructed as something to be accommodated (a much better move than something to be ignored), but it is not yet seen as "enabling insight—critical, experiential, cognitive, and sensory" (Brueggemann, 2002, p. 121). Although

Ray and Ray provided very valuable information on accessible design, they did not imagine the possibility that technical communicators with disabilities may have embodied knowledges that could transform and improve Web design practice for all users.

Furthermore Ray and Ray's (1998) equation of accessibility with adaptive technologies—a common move in most articles of this type—also inadvertently reinforces the ideology of normalcy. We technical communicators must question the implications of how the adaptive technologies used by people with disabilities are implicitly defined in opposition to the technologies (i.e., market-leading Web browsers) used by everyone else. After all, dominant technologies are adaptive as well; the problem is that they are adapted to "normal" users and thus exclude many other people. Rather than just ensuring that our texts can be read by adaptive technologies, we should begin to argue for deep structural changes that could make all technologies more accessible and that could provide all people with more choices for accessing content. For example, although screen (text-to-speech) readers are an invaluable tool for many users with visual impairments or blindness, we must question why so many websites present content only in text format in the first place. Rather than putting the onus on the user to employ adaptive technology to turn text to speech, we might provide audio, textual, and visual versions of Web content and let all users—even those who do not identify as having a disability—choose the format that is best for them.

Indeed many so-called adaptive technologies can ultimately open up communication possibilities for wide varieties of people. For example, one of the first typewriters was developed as an assistive technology for the blind (Adler, 1973), and one of the earliest forms of synchronous electronic conferencing (ENFI) was developed for deaf and hard-of-hearing students at Gallaudet University (Barclay, 1995). Subverting the opposition between normal and "adaptive" technologies, we can begin to open up a space in which the embodied experiences and knowledges of users with disabilities provide insights for improving communication practices for all people.

Enabling/Normalizing Participatory Design

Salvo's (2001) "Ethics of Engagement: User-Centered Design and Rhetorical Methodology" came much closer to advocating a model of usability based on the notion of disability as enabling insight. In this article, Salvo drew upon Bakhtinian discourse theory and participatory (or Scandinavian) design practices to argue that designers should adopt a dialogic ethic of collaboratively developing products with users in local contextual environments. As one of three examples of this kind of practice, Salvo discussed Whitehouse's (1999) work of collaboratively designing tactile signage with users of the Lighthouse, "an office designed to serve the needs of the blind and sight impaired. ...The first insight the users provided was that many of the users of the Lighthouse office space do not read Braille. ...Users had loss of sight in common, but very little else, being of all ages, socioeconomic backgrounds, levels of experience with hearing and touch to compensate for sight loss. In this case, scientific research methods did not apply. What allowed the designers to create useful information-rich products for this situation was their interaction and communication with their clients" (p. 283).

Salvo (2001) discussed a compelling example of ways in which the involvement of people with disabilities in design processes can be transformative. Yet, there is something potentially troubling in how the Lighthouse story of accessible design is positioned in the article as a whole. The other two case studies that Salvo cited do not address disability in any specific way. Furthermore, Salvo did not include disability studies scholarship in outlining theoretical support for participatory design practices. As Salvo was ultimately arguing for a model of design that is highly local and contextual, he ended up implicitly suggesting that disability as insight most applies in institutional spaces in which people with disabilities are prominent. As many local contexts in the world today exclude people with disabilities, the progressive potential of collaborating with users may have limits. If all the potential local users of a technology harbor unconscious ableist biases, they might create a usable yet exclusionary technology. In other words, although participatory design can be great model for reconceiving of usability in ways that value the perspectives of people with disabilities, a disability studies cultural critique demands that we (at times) challenge the assumptions of users in normalized contexts.

Ultimately, technical communicators must move from seeing disability accessibility as a concern particular to a subset of users and begin to reimagine it as a source of transformative insight into design practice for all. In the following section on pedagogical interventions, I offer numerous tentative suggestions for how to begin this process.

DISRUPTING NORMALCY: PEDAGOGICAL INTERVENTIONS

Within technical communication, cultural studies scholars have developed pedagogies that blend critique and intervention. Arguing that scholars must teach students to critically analyze the social implications of how technical communication texts are produced, distributed, interpreted, and consumed, Scott (2004) offered a pedagogy that encourages students to "become more aware and critical of the ways texts are transformed, more aware of the broader conditions shaping them, and more attentive to their multiple effects, especially on users. But this critical awareness can only be the beginning of a cultural studies approach; it must be accompanied by the impulse to respond ethically to the problematic functions and effects of texts" (Scott, 2004, p. 215).

Extending this cultural studies pedagogy, disability studies calls for teaching students to interrogate and to intervene in the ways in which technical communication practices work to reinforce ableist hierarchies. To this end, I offer some suggestions for how disability studies could inform instruction on communication technology, usability, linguistic bias, narrative, and discourse communities.

Rethinking Discourses of Assistive Technology and Usability

When teaching students about Web design, technical communication pedagogues must continue to provide instruction in adhering to Web accessibility standards. Often, in discussing standards, we technical communicators focus on ensuring compatibility with assistive (or adaptive) technologies such as screen readers. Although teaching about assistive technology and Web standards is an essential step in increasing access, we must begin to trouble the binary between normal and assistive technologies. Challenging the naturalization of conventional ableist technologies, we should teach students to view all technologies as assistive. For example, both Internet Explorer and the JAWS Screen Reader can be seen as technologies that assist users in reading Hypertext Markup Language (HTML) code.

Too often in our computer classrooms, so-called assistive technologies are segregated. We may have only one station with screen-reading or text-zooming software, a station that is reserved for users who are blind or visually impaired, or for testing of websites for accessibility. Video and multimedia texts often are captioned only at the request of individuals who are deaf or hard of hearing. Speech-to-text composing software is also often similarly segregated. Challenging this marginalization of technologies used by people with disabilities, we should create pedagogical environments in which all students use and critically analyze screen-reading, captioning, and speech-to-text technologies. The point of these activities would not be to give students the experience of being a disabled user (a largely impossible and questionable goal), but to get students to see how the experience of using various assistive communication technologies can help open up possibilities for reimagining communication design. (Students who were already regular users of these technologies would also gain the opportunity to reflect on them critically.)

For example, students could conduct a contrastive analysis of their experiences reading a website with a conventional Web browser and with a screen reader. Some questions students could consider include

1. How does your understanding of the website differ in the two assistive media? What elements of the website were emphasized and de-emphasized in each of the media?

2. What are the potential advantages of hearing a website? What kind of possibilities does this open up?

3. How would website design principles change if listening to a website was the dominant mode of communication?

4. How might a website be redesigned to provide options for users to choose to audio, visual, or textual versions of content?

In this way, students would come to see assistive technologies as sources of insight into design processes rather than just as marginal tools to be accommodated. This activity could also benefit teachers in rethinking their own pedagogical websites (ideally with student input). In addition to looking at Web technologies, students could also consider examples of common household products. By having students read a case study of the development of assistive household utensils (Center for Universal Design, 2000), teachers could demonstrate how many of the revolutions in utensil design (e.g., the development of more comfortable, larger grips and more visible and tactile markings) initially began as assistive technologies for individuals with mobility and visual impairments. For a kinesthetic learning experience, teachers could have students actually use the old and new versions of the utensils, coming to recognize how these assistive technologies ultimately made such activities as measuring and vegetable peeling easier for everyone.

In addition to critically interrogating assistive technology discourses in which people with disabilities are highly present (albeit in problematic ways), we technical communication scholars must also critically intervene in broader usability discourses in which people with disabilities are often absent or marginalized. Most discussions of usability methodologies center on how much input the user has in the design of the document or of the technological product; traditional models of usability testing often involve users after a product or document has been developed, whereas other more recent approaches such as participatory design and contextual inquiry involve users much more intimately in development processes. Yet in teaching students to evaluate usability methodologies, we must do more than teach them to interrogate users' levels of participation and power; rather, we must teach students to question how definitions or selections of average users ultimately may work to reinforce the ideology of normalcy.

Some questions that could guide critical discussion and critical reflective writing about usability include:

1. When technical communicators conduct a task analysis to prepare to write a manual, what kinds of assumptions do they make about the abilities of the users who will have to perform the task?

2. When usability tests are conducted to find the most common problems users would have with a document or product, to what extent are the concerns of users with disabilities addressed and given prominence?

3. If technical communicators work on a participatory design project with users in an office that is inaccessible for people with disabilities, will they merely reinforce the ableist biases of the users they are serving?
4. How could the perspectives of users with disabilities ultimately lead to design improvements that can benefit all users?

In addition to having students reflect critically on these issues, both cultural studies and disability studies approaches demand that students and teachers consider ways to intervene in social constructions of assistive technology and usability. For a start, students and teachers could work together to begin to transform normalizing technological environments on campus. Working collaboratively, students could research and draft persuasive proposals about how campus courseware programs could be made more accessible to users with disabilities and about how assistive technologies should be more widely available to all students. Extending this work, students and teachers could collaboratively create course materials that enable all users to benefit from choices in how information is presented (visually, textually, aurally, and kinesthetically).

Critiquing Linguistic Ableism

As a result of feminist interventions (Sauer, 1993, 1994; Vaugh, 1989), we technical communication teachers now commonly teach students to consider the gendered power dynamics of language use and to avoid explicitly sexist language practices (generic masculine pronouns, sexist metaphors). Extending feminist pedagogical perspectives on the relationship between language and power, disability studies asks us to uncover and challenge the pervasiveness of linguistic ableism (Linton, 1998), that is, to consider how the conventions of the English language ultimately work to reinforce the ideology of normalcy. To give students a good introduction to the political implications of naming disability, I recommend assigning and discussing Linton's (1998) highly readable chapter "Reassigning Meaning" (pp. 9–33). In this chapter, Linton reconceptualized disability as a social/political identity rather than a medical designation. She then critiqued numerous naming practices that work to restrict social/political agency of people with disabilities: pejorative attacking words (*cripple, deformed, maimed, freak*); condescending words (*challenged, special*); passive, victim words (*confined to a wheelchair, stricken with*). In addition, Linton critiqued the common tendency for people who do not have disabilities to speak of disability in terms of overcoming. "The popular phrase overcoming a disability is used most often to describe someone with a disability who seems competent and successful in some way. This idea is reinforced by the equally confounding statement 'I never think of you as disabled.' An implication of these statements is that the other members of the group from which the individual has supposedly moved beyond are not as brave, strong, or extraordinary as the person who has 'overcome' that designation" (Linton, pp. 17–18).

Many well-meaning students might not see the problem with complimenting people for rising above their disabilities and therefore might later draw on this trope in professional documents such as performance appraisals and recommendation letters. Through reading and discussing Linton's (1998) text, students can begin to interrogate the subtly ableist implications of overcoming rhetoric, noting ways in which this rhetoric impedes the kinds of collective social change necessarily to address ableist oppression.

Yet this discussion of naming disability is only the beginning of an interrogation of linguistic ableism. Drawing implicity on Mitchell's (2002) work on the material implications of disability metaphor, I ask students to begin listing metaphors and expressions that reference disability. Some examples include *blind, blindspot, fall on deaf ears, dumbing down, crazy*, and *idiot*.[4] After generating a substantial list of metaphors with the class, teachers can then ask students to analyze them. Why is it that all (or most) of the metaphors that use disability carry negative connotations? What do these metaphors reveal about how our culture views disability? How might these metaphors help construct the reality of experiencing disability in our culture? By doing this activity early in the course, teachers can encourage students to interrogate disability metaphors in all the texts that they read and write.

Incorporating Disability Autobiography

In the past few years, scholars have argued against the traditional devaluing and exclusion of narrative texts from professional communication study. As Perkins and Blyler (1999) demonstrated, the failure to grant narrative serious scholarly attention has been accompanied by the exclusion of a range of topics from consideration. For example, feminist scholars have asserted that, "as narrative ways of knowing have been marginalized, women's experiences and some types of discourse often produced by women have suffered neglect as well" (p. 21).

Extending this discussion of the politics of narrative, I suggest that a disability studies pedagogy also necessitates an inclusion of heretofore excluded narrative texts into the technical communication classroom. If we limit ourselves to analyzing technical and professional (legal, medical, rehabilitative) texts about disability, we will end up reinforcing the normalizing construction of disability as an individ-

---

[4]This discussion of linguistic ableism is particularly pressing because metaphoric uses of disability are actually quite common in the technical communication scholarly literature. Some examples include *dumbing down* (Mazur, 2000; Kostelnick, 1998), *idiot* (Johnson, 1998), *blind yourself* (Hart, 2000), *blinding influence* (Dragga & Voss, 2003), and *blind spots* (Uljin & St. Amant, 2000).

ual problem or case to be fixed, thereby deadening the agency of those with disabilities who seek to define their own needs and interests.

Although one could include a number of texts to accomplish this, I suggest a recent edited collection as particularly salient to the workplace focus of most technical communication classes. O'Brien (2004) juxtaposed numerous fictional and autobiographic stories of disability-based discrimination faced by employees and patrons of public and private workplaces along with legal commentary about how these narratives might be viewed in terms of recent ADA case law. This tactic of interspersing commentary with narrative effectively highlights the limiting ways in which ADA case law has defined disability and the lack of agency it affords people with disabilities who seek to articulate their own identities and needs. In addition to introducing students to important aspects of disability law and policy that they will need to know on the job, this collection would also encourage students to interrogate the limitations of current disability policies, imagining alternatives to existing practices.

Although I think that autobiographical disability narratives offer an important corrective to ableist professional discourses, technical communication scholars should be careful not to celebrate disability autobiography uncritically as representations of reality. As Mitchell (2000) demonstrated, some disability autobiographies work against a social/political understanding of disability, at times reinforcing problematic tropes of disability experience as an individual problem to be overcome (self-reliance) or an individual tragedy to be pitied (sentimentality). In other words, although autobiographical narratives of disability can offer important insights, they too must be critically interrogated for the material implications of how they construct disability identities.

Challenging Discourse Community Norms

In contrast to conventional technical communication pedagogies, which teach students how to assimilate the norms of professional and organizational discourse communities, cultural studies pedagogies seek to teach students, as Herndl (1993) wrote, to "participate in professional discourse, but also to recognize it as contingent and ideologically interested," thereby enabling students to "escape the 'culture of silence' fostered by an uncritical attitude to hegemonic discourse" (p. 225). Yoking critique and intervention, Henry (2000) had his professional writing students conduct autoethnographic projects in which they both analyzed and challenged the discursive practices that form their subjectivities and that constrain their actions as professional writers. Extending this pedagogical work, disability studies theory can open additional critical questions that students can ask in textual and ethnographic analyses of discourse communities:

1. Are people with diverse abilities visible in the textual and visual representations of the community? If not, why not? If so, what do the representations tell you about how disability is viewed within this community?
2. What forms or modes of communication (visual, print, oral) are privileged and devalued in the community? Who would most likely be included and excluded from participation in the privileged forms of communication?
3. What material and virtual spaces does the community inhabit? For what kinds of bodies were these spaces designed? What bodies are excluded from full participation in these spaces?

Following this kind of analysis, students could then produce a persuasive professional document (memo, proposal, presentation) in which they make specific recommendations about how a particular organization or profession could change its practices to be more inclusive of people with diverse abilities.

Enabling Insights/Transforming Realities

Almost all technical communication practices are embedded in the construction of normalcy in one way or another. Thus integrating disability studies cultural critique is not a matter of adding one assignment or one activity. Rather, just as we integrate audience analysis into almost everything we teach, we should also consider how disability may offer insight into the myriad communication situations we face as teachers, as professionals, and as students. As Davis (1995) noted, "the consideration of disability ... rather than being a marginal and eccentric focus of study, goes to the heart of issues about representation, communication, language, ideology, and so on" (p. 124). Through a sustained engagement with disability studies, technical communicators can continue to reimagine the field in ways that productively contribute to the material social changes necessary to create a more equitable, accessible society.

ACKNOWLEDGMENTS

I would like to thank Brenda Brueggemann, Rita Rich, Kristina Torres, and Rebecca Dingo for many insightful conversations that helped me develop and refine this project. I also would like to thank the editors and reviewers for their thoughtful suggestions.

REFERENCES

Adler, M. H. (1973). *The writing machine*. London: George Allen & Unwin.
Barclay, D. (1995). Ire, envy, irony, and ENFI: Electronic conferences as unreliable narrative. *Computers and Composition, 12*, 23–44.

Brueggemann, B. J. (1999). *Lend me your ear: Rhetorical constructions of deafness.* Washington, D.C.: Gallaudet University Press.
Brueggemann, B. J. (2002) An enabling pedagogy. In S. L. Snyder, B. J. Brueggemann, & R. Garland-Thomson (Eds.), *Disability studies: Enabling the humanities* (pp. 317–336). New York: The Modern Language Association.
Carter, J., & Markel, M. (2001). Web accessibility for people with disabilities: An introduction for web developers. *IEEE Transactions on Professional Communication, 44,* 225–233.
Center for Universal Design. (2000). *Oxo International becomes universal design icon.* Retrieved on January 10, 2005, from http://design.ncsu.edu/cud/proj_services/projects/case_studies/oxo.htm
Colker, R. (1999). The Americans with Disabilities Act: A windfall for defendants. *Harvard Civil Rights-Civil Liberties Law Review, 34,* 98–162.
Davis, L. (1995). *Enforcing normalcy: Disability, deafness, and the body.* New York: Verso.
Dragga, S., & Voss, D. (2003). Hiding humanity: Verbal and visual ethics in accident reports. *Technical Communication, 50,* 61–82.
Eastman, C. (1910). *Work-accidents and the law.* New York: Charities Publication Committee.
Garland-Thomson, R. (1997). *Extraordinary bodies: Figuring physical disability in American culture and literature.* New York: Columbia University Press.
Garland-Thomson, R. (2002). The politics of staring: Visual rhetorics of disability in popular photography. In S. L. Snyder, B. J. Brueggemann, & R. Garland-Thomson (Eds.), *Disability studies: Enabling the humanities* (pp. 56–75). New York: The Modern Language Association.
Hart, G. (2000). Ten technical communication myths. *Technical Communication, 47,* 291–298.
Henry, J. (2000). *Writing workplace cultures: An archaeology of the discourse community.* Carbondale: Southern Illinois University Press.
Herndl, C. G. (1993). Teaching discourse and reproducing culture: A critique of research and pedagogy in professional and non-academic writing. *College Composition and Communication, 44,* 349–363.
Hubbard, R. (1997). Abortion and disability: Who should inhabit the world? In L. J. Davis (Ed.), *Disability studies reader* (pp. 187–202). New York: Routledge & Kegan Paul.
Johnson, R. (1998). *User-centered technology: A rhetorical theory for computers and other mundane artifacts.* Albany: State University of New York Press.
Kolstenick, C. (1998). Conflicting standards for displaying data displays: Following, flouting, and reconciling them. *Technical Communication, 45,* 473–482.
Lay, M. M. (2000). *The rhetoric of midwifery: Gender, knowledge, and power.* New Brunswick, NJ: Rutgers University Press.
Linton, S. (1998). *Claiming disability: Knowledge and identity.* New York: New York University Press.
Longo, B. (1998). An approach for applying cultural study theory to technical writing research. *Technical Communication Quarterly, 7,* 53–73.
Longo, B. (2000). *Spurious coin: A history of science, management, and technical writing.* Albany: State University of New York Press.
Madaus, M. (1997). Women's role in creating the field of health and safety communication. *Technical Communication Quarterly, 6,* 261–280.
Mazur, B. (2000). Revisiting plain language. *Technical Communication, 47,* 205–211.
Mitchell, D. T. (2000). Body solitaire: The singular subject of disability autobiography. *American Quarterly, 52,* 311–315.
Mitchell, D. T. (2002). Narrative prosthesis and the materiality of metaphor. In S. L. Snyder, B. J. Brueggemann, & R. Garland-Thomson (Eds.), *Disability studies: Enabling the humanities* (pp. 15–30). New York: The Modern Language Association.
O'Brien, R. (Ed.). (2004). *Voices from the edge: Narratives of the Americans with Disabilities Act.* New York: Oxford University Press.
O'Hara, K. (2004). "Curb cuts" on the information highway: Older adults and the Internet. *Technical Communication Quarterly, 13,* 423–445.

Perkins, J., & Blyler, N. (1999). Introduction: Taking a narrative turn in professional communication. In N. Blyler & J. Perkins (Eds.), *Narrative and professional communication* (pp. 1–37). Stamford, CT: Ablex.

Ray, D. S., & Ray, E. J. (1998). Adaptive technologies for the visually impaired: The role of technical communicators. *Technical Communication, 45,* 573–579.

Russell, M. (1998). *Beyond ramps: Disability at the end of the social contract.* Monroe, ME: Common Courage Press.

Salvo, M. J. (2001). Ethics of engagement: User-centered design and rhetorical methodology. *Technical Communication Quarterly, 10,* 273–290.

Sauer, B. A. (1993). Sense and sensibility in technical documentation: How feminist interpretation strategies can save lives in the nation's mines. *Journal of Business and Technical Communication, 7,* 63–83.

Sauer, B. A. (1994). Sexual dynamics of the profession: Articulating the *ecriture* masculine of science and technology. *Technical Communication Quarterly, 3,* 309–323.

Schlosser, E. (2002). *Fast food nation: The dark side of the all-American meal.* New York: Houghton Mifflin.

Scott, J. B. (2003). *Risky rhetoric: AIDS and the cultural practices of HIV testing.* Carbondale: Southern Illinois University Press.

Scott, J. B. (2004). Tracking rapid HIV testing through the cultural circuit: Implications for technical communication. *Journal of Technical and Business Communication, 18,* 198–219.

Uljin, J. M., & St. Amant, K. (2000). Mutual intercultural communication: How does it affect technical communication? *Technical Communication, 47,* 220–237.

Vaugh, J. (1989). Sexist language—still flourishing. *Technical Writing Teacher, 16,* 33–40.

Waddell, C. (1990). The role of pathos in the decision-making process: A study of the rhetoric of science policy. *Quarterly Journal of Speech, 76,* 381–400.

Whitehouse, R. (1999). The uniqueness of individual perception. In R. Jacobson (Ed.), *Information design* (pp. 103–129). Cambridge, MA: MIT Press.

Wilson, J. C. (2000). Making disability visible: How disability studies might transform the medical and science writing classroom. *Technical Communication Quarterly, 9,* 149–161.

Wilson, J. C., & Lewiecki-Wilson, C. (Eds.). (2001). *Embodied rhetorics: Disability in language and culture.* Carbondale: Southern Illinois University Press.

Jason Palmeri is a doctoral candidate studying rhetoric, composition, and literacy at Ohio State University. He can be reached at palmeri.2@osu.edu.

Cars, Culture, and Tactical Technical Communication

Miles A. Kimball
Texas Tech University

This article examines two cases of technical documentation occurring outside of institutions. Using a framework derived from de Certeau's (1984) distinction between strategies and tactics and Johnson's (1998) concept of the user-as-producer, I analyze communities surrounding Muir's (1969) *How to Keep Your Volkswagen Alive! A Manual of Step by Step Procedures for the Compleat Idiot* and Champion's (2000) *Build Your Own Sports Car for as Little as £250*. These communities engage in tactical technical communication, especially in the form of technological narratives that participate in broader cultural narratives about technology.

Technical communication has come a long way since its birth as a mechanism for what Longo (2000) called "management system control" (pp. 100–121). But despite the increasing investment in user-centered design, most of what we recognize as technical communication still begins and ends with corporate, governmental, or organizational agendas. As technology becomes further integrated into our daily lives, however, we should expect to see not only more complex user interactions involving institutional technical communication, but also more technical communication happening outside, between, and through corporations and other institutions. Understanding this development is an essential step in conceptualizing a technical communication for a postindustrial world.

In this article, I examine two cases of extra-institutional technical communication: the documentary cultures surrounding Muir's (1969) *How to Keep Your Volkswagen Alive! A Manual of Step by Step Procedures for the Compleat Idiot* and Champion's (2000) *Build Your Own Sports Car for as Little as £250*. Despite and perhaps because of their mundane and extracorporate status, these documents have been influential in creating and shaping cultures, especially those surrounding the automobile. Both books participate in a technological narrative of the self-sufficient technologist—a person who counters a feeling of helplessness in a dominant culture by living as an independent operator, a technological scavenger on the periphery of industrial society.

In examining these cases, I follow the lead of scholars such as Longo (1998) who have adapted cultural studies approaches to put forward a broader conception of technical communication as a human activity happening both within institutions (or organizations, as we in technical communication more often refer to them) and in the gaps between them. In addition, by analyzing the dynamics of two cases, one before and one after the rise of the Internet, I outline how users have appropriated technology to increase their freedom of agency and their involvement in shared cultural narratives about technology, as members of these cultural groups form communities to create and share their own technical documentation. Rather than working as designers forming user-centered technologies and documentation, these people combine the positions of the user and the designer, re-creating technology for their own purposes.

BROADENING THE ORGANIZATIONAL FOCUS IN TECHNICAL COMMUNICATION

Harrison and Katz (1997) commented, "Although writing obviously takes place in other nonacademic contexts, many would acknowledge that organizations are the most frequent social context in which technical communication takes place" (p. 18). Despite their careful qualifications ("Although," "many," "most frequent"), Harrison and Katz's main point was that technical communication should, indeed, focus on organizations—and rightly so, given that organizations are important settings for technical communication. In keeping with this focus, technical communication scholars, teachers, and practitioners have made valuable contributions to understanding how organizations communicate. But perhaps our focus on the organization has kept us from appreciating the growing amount of technical communication produced outside of (or in spite of) organizations.

The power of this organizational focus in technical communication is considerable. In pedagogy, textbooks uniformly introduce technical writing as a workplace skill. For example, Houp, Pearsall, Tebeaux, and Dragga's (2002) textbook begins its section headed "The Substance of Technical Writing" with, "Organizations produce technical writing for internal and external use" (p. 2), assuming from the outset that technical writing is an organizational practice. And Markel's (2004) textbook defined technical communication with "seven major characteristics," one being that it "reflects an organization's goals and culture" (p. 7). In pedagogical scholarship, we repeatedly debate the best practices of teaching students to fit into institutional roles as technical communicators (see for example, Harrison & Katz, 1997; Meyer & Bernhardt, 1997). The organizational focus has influenced even pedagogies trying to broaden the scope of technical communication, such as service learning. As Ornatowski and Bekins (2004) pointed out, the pedagogical justification for sending students to work with community organizations assumes that students' experiences there will translate to professional skills that may be used in corporate jobs (p. 255). Scott (2004) similarly noted that service-learning has not always lived up to its potential for promoting civic engagement, focusing instead on a narrow pragmatism of "institutional practices and structures" (p. 289). This focus applies to technical communication scholarship as well, where examples of research at institutional sites are legion. Even research approaches that might examine documentary cultures outside of institutions have thus far focused on institutions. Social construction, for example, holds promise for examining how something as organic as a whole society communicates about technology. But much of the social construction research assumes the organization as the social unit most worthy of study: Subbiah (1997) summarized social construction as an approach that "assumes that knowledge is constructed in communities by members who share beliefs, language, practices, and values"; as such, it focuses on the " 'atmosphere,' 'company spirit,' and 'ethos' " of "organizational culture" (pp. 58, 59).

This focus on the organization is worthwhile; we should research institutional cultures and communication practices because corporations and governments establish the landscape of power in postindustrial societies. But as other scholars have pointed out, we also need to broaden our field of view to account for technical communication as a practice extending beyond and between organizations. A number of scholars have criticized the field's focus on meeting corporations' communication and job-training needs. Bushnell (1999), for example, complained that technical communication pedagogy has "generally accepted and, worse, *internalized* corporate paradigms," with academic programs acting as a "training department for students' future employers" (p. 175). And Johnson-Eilola (2004) argued that technical communication risks demotion to a service industry, when instead technical communicators should be doing "symbolic-analytic work" (p. 177).

Others have extended the scope of our scholarship to contexts broader than organizations. Russell (1997) suggested that people use genres as conventionalized strategies to negotiate a variety of activity systems—not just the unitary scientific laboratory, to use his example, but the web of activity systems that touch on that site: scientific disciplines, the university that built the facilities, agencies funding the research, students who work or study in the lab, and families who pay their tuition. As Feenberg (1999) pointed out, "social systems are very much in the eye of the beholder," their boundaries depending on the perception of those within or outside of the system; instead of monolithic hegemonies, we live among competing, overlapping " 'networks' of systems" (p. 118). Johnson-Eilola (2005) posited an already-existing world in which people dynamically gather and create data with which to live their lives in a variety of organizational and nonorganizational settings. A growing number of scholars have also called for communitarian approaches to understanding technology and technical communication. For example, Grabill (2003) and Harrison, Zappen, and Prell (2002) examined community websites as spaces for information exchange. And recent *Technical Communication*

Quarterly special issues on civic engagement (Carpenter & Dubinsky, 2004) and on public-policy writing (Rude, 2000) have attested to a growing desire to broaden and redirect scholars' focus from organizations exclusively to the broader social and cultural landscape between and across (and thus including) organizations.

This article seeks to continue this broadening of technical communication, specifically by focusing on occasions where users of technical communication become producers of technical communication in settings not defined by institutions. In doing so, I extend Johnson's (1998) concept of the user-as-producer. Johnson proposed three common concepts of users: *users-as-practitioners*, mere tool-users or so-called "idiots" (although with situational cunning, or *metis*); *users-as-producers* of knowledge derived from their experience; and *users-as-citizens*, "active, responsible members of the technological community" whom designers should ask to participate in technological design (p. 61). Johnson suggested that designers typically think of users primarily in the first role, as tool-users to whom technology must be explained. Instead, Johnson contended, designers should value users by placing them at the center of technological development, incorporating them as full participants in every aspect of the design process: "The rhetorical situation involving users, designers, and artifacts should interact in a negotiated manner so that technological development, dissemination, and use are accomplished through an egalitarian process that has its end in the user" (p. 85).

This formulation of user-centered design positively focuses the design process on users, but it also retains a certain separation between users and designers. The designers Johnson (1998) described are typically different people than the users— for the most part the designers represent institutions (such as corporations and governments), and thus need to seek out user participation in design. Citing the ancient Greeks' example of Athena as the builder as well as the driver of her chariot, Johnson did hold out the possibility of users becoming producers not only of knowledge but also of their own physical products (p. 58). But he stopped short of seeing that role as a current reality:

> We have become a culture that wears its labels on the outside of its clothing, and we seem quite comfortable with that display of end-use consumption. We seem comfortable only driving, not building and driving. ... Can a refigured sense of what it means to *use* really create changes in our basic culture? We might try, and as a result we could very possibly *reinvent*, users' ways of knowing. People as producers, people as participants, not as idiots. Imagine. (p. 67)

This article analyzes the work of people engaged in both "building and driving" the chariot of our day: the automobile. Muir's Volkswagen manual and Champion's sports car build-manual marked the development of communities of independent user-designers who produce not just valuable tacit or suppressed knowledge, but actual documentation of their technical practices and experiences, especially in the form of technological narratives. In the case of the community surrounding Champion's book, they actually build and drive the car of their dreams. These user-producers engage in productive acts outside of and, indeed, in resistance to institutional agendas.

Understanding these user-producers, however, requires a theoretical framework that foregrounds the practices of users as they work between and across institutions, seeking to find their own ways to their own ends while negotiating the contested "'networks' of systems" Feenberg (1999, p. 118) described. In the next section, I sketch one approach that gives a good framework for analyzing user-producer communities: de Certeau's (1984) distinction between strategies and tactics.

INSTITUTIONAL STRATEGIES, INDIVIDUAL TACTICS

According to de Certeau (1984), strategies are systems, plans of action, narratives, and designs created by institutions to influence, guide, and at worst manipulate human society. De Certeau often used the example of the city as a strategic system—a fabric of designs laid down by planners (streets, zones), producers (factories, stores, markets), and governments (rules, laws, regulations). Strategies are thus written onto the social landscape, forming the rules of individual action.

If institutions own the house and set up the game strategically, individual players try to overcome the odds with good tactics. Individuals use tactics to survive and to come as close to achieving their purposes as possible. A tactic differs from a strategy in that it lacks a place or "property," relying instead on action in the "cracks" between "proprietary powers": "It poaches in them. It creates surprises in them. It can be where it is least expected. It is a guileful ruse. In short, a tactic is an art of the weak" (de Certeau, 1984, p. 37). For example, a traveler could employ tactics to navigate a city. Although constrained by the strategies of place—the layout of streets; the rules of the road; private properties; and economic, cultural, or governmental spaces—the traveler could still make choices about how to get to the destination and fulfill personal agendas: "I'll go this way, it's prettier"; "I'll go that way so I can pick up some bread for dinner." By walking (or driving) through a city of possible routes, the traveler, according to de Certeau (1984), also creates the city by "actualizing some of these possibilities" and increasing their number "by creating shortcuts and detours" (p. 98). Although the strategic world values commerce, regularity of order, and power derived from the authority of place, tactics focus on, in de Certeau's words, "an economy of the '*gift*' ... , an esthetics of '*tricks*' ... and an ethics of *tenacity*" (p. 26).

De Certeau (1984) focused on two tactics: *bricolage* and *la perruque*. Bricolage includes practices in which the individual (the *bricoleur*) puts together cultural ingredients creatively for his or her own purposes. Bricolage is often translated as "making-do," a meaty pun on this combination of making and doing what you can with what you have. Bricolage involves a clever, witty, tricky, even illicit appropriation (de Certeau, 1984, p. xii). Further, bricolage as a creative act transforms consumers into producers by making objects do something other than what they were designed to do (p. xii).

To bricolage, de Certeau (1984) added *la perruque*, a tactic arising from the workplace. A French euphemism literally meaning "the wig," la perruque involves appropriating time or surplus material at work to personal uses. According to de Certeau, examples include "a secretary's writing a love letter on 'company time' or ... a cabinetmaker's 'borrowing' a lathe to make a piece of furniture for his living room" (p. 25). This appropriation typically involves a pleasure in making something for its own sake, to show the worker's skill and craftsmanship (p. 25). Although retrograde, this practice, de Certeau pointed out, is not unsophisticated: "Far from being a regression toward a mode of production organized around artisans or individuals, la perruque re-introduces 'popular' techniques of other times and other places into the industrial space" (p. 25). Despite the essentially illicit nature of la perruque, employers often turn a blind eye to the practice, allowing it to continue as long as the infractions are not severe.

Given the extension of work into all areas of people's lives in today's wired and wireless world, it is not surprising that bricolage and la perruque have become significant concepts for discussing the ways consumers appropriate things outside of the standard, one-way path of capitalistic buying and consuming of goods and services. These dynamic acts of cultural appropriation, consumption, and production, of making something new out of old materials or using old techniques in new settings, finds many expressions in today's world: artists who engage in collage with found objects, musicians who sample old songs to make new ones, Web designers who appropriate the Hypertext Markup Language (HTML) <table> tagset to use for page layout, and hobbyists who hack the firmware of their cars for greater efficiency or performance. Bricolage and la perruque become important ways of making do in a postindustrial world.

These tactics, according to de Certeau (1984), operate in material and in linguistic and rhetorical terms. Strategies are to tactics as the rules of language (*langue*) are to the possibilities of individual expression (*parole*; de Certeau, 1984, pp. 32–33). Strategies are the rules of logic, whereas tactics are the persuasions of sophism, temporarily "mak[ing] the weaker position seem the stronger" (p. xx). Strategies depend on owning the text; tactics depend on propitious timing (*kairos*) and situational cunning (*metis*) in using and making the text. A reader even uses tactics to remake the text: The reader "insinuates into another person's text the ruses of pleasure and appropriation: he poaches on it, is transported into it, pluralizes himself in it [...] . A different world (the reader's) slips into the author's place. This mutation makes the text habitable, like a rented apartment" (p. xxi). Readers are therefore involved in a creative, tactical "consumer production" of the text. This creative activity links the material object and the textual object; the individual writes the object as much as he or she writes a narrative about and through it. Both material and textual objects thus become texts written and rewritten by the tactician.

TECHNICAL COMMUNICATION AND TACTICAL NARRATIVES

The significance of de Certeau's (1984) ideas for technical communication arises from this link of objects to texts, and more specifically, to narrative texts. As the industrial revolution mechanized technique, removing it from the hands of craftsmen and embodying it in machines controlled by engineers, the "know-how" of craftsmen became separated from descriptive accounts (such as Diderot's *Encyclopedie*) that had "objectively articulated it with respect to a 'how-to-do' " (de Certeau, 1984, p. 69). Formal technical documentation developed as a situated strategy, separate from the tactical practices of workers—or in Johnson's (1998) terms, designers became separated from users. Workers relied more heavily on a "*narrativity* for everyday practices," expressing their tactics "not only in daily practices [...] but also in rambling, wily, everyday stories" (de Certeau, pp. 70, 89). This narrativity allows the prospect of users not only producing and engaging in tactics, but also sharing them through tales of fooling, tricking, and taking advantage of the strategic system.

The links between technical communication and narrative have been explored thoroughly, most obviously in Barton and Barton's (1988) groundbreaking article titled "Narration in Technical Communication" and in Blyler and Perkins' (1999) collection *Narrative and Professional Communication*. Barton and Barton pointed out that technical texts can be shaped by narrative on several levels, from "master narratives" such as the scientific method, to the technical narratives of manuals, process descriptions, and procedures, and even to the narratives of objects, which have an implicit "story" of use (pp. 40–43). At minimum, technical documents participate in two narrative levels, communal and local. Technical documents build from and contribute to communal narratives and cultural myths. As Bushnell (1999) argued, "Technology and science are not neutral but function within social, political, historical, and cultural contexts or 'narratives'; because of this, technical and scientific communication can be expected not only to reflect but to help *create* those narratives" (p. 181). Even traditional technical documents embody local technological narratives of strategies for work and actions by telling ideal narratives of how machines, technologies, and processes should work (manuals, procedures), could work (proposals), or did work (reports).

The tactical narratives suggested by de Certeau (1984) would bridge these two levels of technological narrative. On the communal level, de Certeau imagined the operation of a larger "communal memory," a shared narrative of tactical resistance to strategic power (p. 87). On the local level, tactical stories differ from the strategic, ideal narratives of traditional technical documentation by telling an individual tactical narrative: not "Here's how it is/was done" but "Here's how I did it." So in examining the technical communication of tactics, one should expect to see less of the traditional genres represented in technical communication textbooks and more of the kind of technological narratives in which users describe their own implementation of tactics, their own version of the shared tactical story. In addition, as de Certeau pointed out, these tactical stories (both local and communal) are typically more implicit than are the strategic stories with which we technical communication scholars are more familiar. Institutions own strategic technical communication and have the power to regulate it as they like. But a tactical technical communication would be more likely under the table, fitting with the trick or surprise of tactical activity.

One promising site for researching the tactical technical communication of user-producers is enthusiast publications and their surrounding cultures. Such publications are motivated not by institutional strategies, but by enthusiasm for the activity, technology, and embedded communal narrative of the subject matter. Researchers could examine any number of enthusiast subjects in this light. In the world of computer gaming, for example, tactics often are shared by users in ad hoc online collections of tricks, tips, and cheats—myriad accounts of ways a user can make the institutionally designed world of the game his or her own, if only temporarily. Of course, many blogs focus on telling such technological narratives, and the new magazine *Make: Technology on Your Time* (2005) is advertised as offering "Dozens of hacks and how-tos for your gear" (cover).

The sections that follow examine two nodes of tactical documentation in automotive enthusiasm. For each example, I analyze first how seminal technical documents (manuals by Muir and Champion) share local tactical advice and technological narratives. Then I discuss how they encourage further tactical actions and documentation on the part of users—primarily as further technological narratives, but also in the second case as formal technical documentation and as technological artifacts—thus exemplifying the combined user-producer about whom Johnson (1998) speculated. Finally, I examine how these cases take part in and contribute to wider cultural narratives of tactical resistance to authority.

TACTICS AND YOUR VOLKSWAGEN

First published in 1969 and now in its 19th edition, Muir's *How to Keep Your Volkswagen Alive! A Manual of Step by Step Procedures for the Compleat Idiot* remains legendary among Volkswagen (VW) Beetle owners. It has sold 2.3 million copies thus far—about one copy for every four Volkswagens ever sold in the United States until 1978, the last year for U.S. Beetle sales (Avalon Travel Publishing, n.d.). As such, the book has paralleled the popularity of the Beetle itself, for a long time the most widely produced vehicle on the planet.

How to Keep Your Volkswagen Alive! differed from most strategic technical documents, which typically arise from institutions and their agendas, in that Muir self-published the book until his death in 1977, making the publication of the book itself a tactical act. This independence allowed Muir to introduce users to Volkswagen repair and maintenance from a stance of subjective resistance, encouraging them to use local tactics to make do with technology. Accordingly, the manual avoided a traditional anonymous or institutional viewpoint, speaking instead through the author's subjective narratives and idiosyncratic voice.

Sharing Local Tactical Advice and Technological Narratives

This subjectivity suggests that Muir's manual fits the tactical patterns described by de Certeau (1984), both in the tactics it suggested to readers and in its own textual practices, which depended heavily on technological narratives. Throughout the manual, Muir treated maintaining a Beetle as a tactical compromise between what the user can provide—labor—and what might be in short supply due to institutional commitments and limitations—time and money. For example, Muir described two possible tactics for major overhaul work, the "two extremes to the 'How To Run a Car Theory.'... One end is to ... take care of breakdowns as they happen.... The other end ... is to keep the car in perfect tune" (p. 154). Muir's choice of tactic depended on local conditions, because "there is bread to consider and sometimes the availability of parts, like a stuck valve twenty miles west of Yuma" (p. 154). This approach differed from traditional automotive manuals by presenting not just ideal procedures, but tactical approaches of making do with what's available—the essence of bricolage. Muir's approach also involved active resistance to and subversion of systems (la perruque). Rather than tell users how to adjust the automatic choke, Muir advised disarming it (adding a comment on his own tactical status): "I refuse to tell you how to adjust it, but I will tell you how to make it *not* work. Do you suppose this is being an activistic reactionary?" (p. 99). Similarly, Muir called the PICT–30 carburetor "another one of those engineering dreams that went wrong" and tells readers how to subvert its pesky electromagnetic cut-off jet (p. 99). In this way, Muir helped readers appropriate control over technology by subverting the institutional practices embodied by technological artifacts, substituting lo-fi techniques to avoid the controls embodied in higher technologies.

To convey this tactical advice, Muir incorporated frequent local technological narratives of tactics undertaken by him or by those who user-tested his manual. He told the stories of Tosh Gregg, who made his way through an engine rebuild by making friends with a good machinist named Elmer, and of John Counter, "who'd never had a car tool in his hands before," but who rebuilt his engine in two weeks of afternoons after his day job (p. 157). Muir also added his own stories. For example, Muir recounted burning a rod in Santa Barbara; after ascertaining the problem, he "ran the cooled engine just enough to get off the freeway and in front of a millionaire's house who let us use the phone ... everyone was wonderful—you know, people are great!" (p. 243).

These stories emphasized the user-mechanic's ability to manage local conditions (including community connections) to tactical advantage. They also humanized technology and situated the reader in a local, personal relationship with the car as its maintainer and master, rather than in an institutional relationship making the reader merely an operator—Johnson's (1998) user-as-practitioner. Muir (1969) suggested an intimate connection between the owner and the car that begins with communication and leads to identification:

> "Come to kindly terms with your Ass for it bears you." Your Volkswagen is not a donkey, but the communication considerations are similar. Your car is constantly telling your senses where it's at: what it's doing and what it needs. ... Talk to the car, then shut up and listen. (p. 3)

The reader's relationship with the car as patterned by Muir's manual gives the car life, or more tellingly, integrates the mechanical life of the car with the biological life of its owner: "Feel with your car; use all your receptive senses and when you find out what it needs, seek the operation out and perform it with love.... Its Karma depends on your desire to make and keep it—ALIVE!" (p. 3). The shared life of the owner with the car, then, becomes an issue of life practices—in this case, one of living slowly, cheaply, economically, and self-sufficiently with technology.

Encouraging User-Producer Documentation

A personal relationship to technology also fostered further technological narratives created by readers, people who used the manual to keep their Volkswagens alive and therefore made the manual a cultural phenomenon. At the Amazon.com (n.d.) website, many of the 58 customer reviews for Muir's manual offer examples of such narratives: "It all began in 1970, with my first VW, a 1967 Squareback"; "I first bought this book back in the 70s, when I think Muir was still alive. It completely captured the hippie, VW era, of which I was a part of, in Big Sur, CA"; "I liked it so much that I bought a VW." This tradition of technological narrative is celebrated in many books that collect Volkswagen owners' anecdotes and memories, of which several are currently or recently in print (see, for example, Rosen, 1999).

These personal anecdotes of life with a car repair manual suggest a deep integration of identity with technology (and with communication about technology); they also illustrate how the readers of Muir's manual and the drivers of Volkswagen Beetles transformed themselves into producers of culture. Muir offered his manual as a chance to keep a Volkswagen alive, but readers' narratives about the manual have kept it in production long after the production of the Beetle ceased. In this way, the narratives of readers, Muir's manual, and the Beetle itself, considered both materially and culturally, have become a shared text that continues to generate additional texts. The users, in de Certeau's (1984) sense, have inhabited the car/text and remade it in their own image.

Contributing to Wider Cultural Narratives of Tactical Resistance

Such reappropriation of a text by consumers paralleled the broader reappropriation of the Beetle's image from its fascist origins to a symbol of 1960s and 1970s counter-cultural resistance. The Beetle was originally produced by Volkswagenwerk AG ("People's Car Works Company"), a corporation Hitler created on Fordist principles to make cars for German workers. The car was originally named the *Kraft-durch-Freude Wagen* ("Strength through Joy Car"), but only a few hundred were manufactured before World War II began, and Volkswagen's production shifted to military vehicles; not one car actually went to workers. Reappropriated after the war from its fascist and Fordist origins, however, the Beetle came to embody resistance to authority, as Beetle owners tactically modified their machines in unique and creative ways. For many years Beetles were likely to be decorated with homemade paint jobs (often brushed), plastered with stickers, festooned with fanciful parts like ears and tails, modified with tongue-in-cheek Rolls-Royce hoods, or transformed into dune buggies. In this regard, the culture surrounding the Beetle appropriated a technology designed to support a fascist state and later a corporation, using it to express a vision for automotive freedom and individuality.

As a text promoting tactical appropriation, Muir's (1969) manual also reflected the broader counter-cultural narrative of its era, which combined contradictory desires for freedom of movement (which requires technology) and for returning to a simpler way of life. This sensibility, often thought of as simply antitechnological, is perhaps better understood as a tactical response to the strategies of military, corporate, and industrial technocracy. In *Questioning Technology*, Feenberg (1999) described the May 1968 student revolts in Paris as a revolt not so much against technology but against technocracy—the bureaucratic (i.e., strategic) control of technology (pp. 21–43). Muir (1973) addressed this distinction in his final book, *The Velvet Monkey Wrench*:

> Machines and computers are our tools. We cannot dis-invent them …. Without them we'd be living a much simpler, day-to-day life, which is a groove, if that's what you choose, but for me it's limiting …. The change needed is to admit we're committed to the machine, then resolve to what degree. (p. 10)

The specific advantage of the Volkswagen was its perceived authenticity in an automotive market overwhelmed with high-tech, jet-inspired tailfins and chrome; in the words of *Popular Mechanics* editor Arthur Railton, "The Volkswagen sells because it is … an honest car. … Wherever [the owner] looks, he sees honest design and workmanship" (Nelson, 1965, p. 219). Thus the Beetle itself is revealed as a perruque—a reintroduction of craft into the industrial system.

LOCOST: A SPORTS CAR FOR THE MASSES

If the Volkswagen community surrounding Muir's manual tactically reappropriated the products of Fordist technocracy, Champion's *How to Build Your Own Sports Car for as Little as £250* established a post-Fordist, postindustrial expression of a similar cultural phenomenon. Of the many automotive how-to books on the market today, few have attracted such a large following. Across the globe, thousands of readers—almost exclusively men— not only have built their own cars, but have also used the Internet to create technical documentation that reflects user-producer tactics. The community surrounding Champion's manual are interested in both building and driving; quintessential user-producers, they have created both original technical documentation (especially in the form of technological narratives) and actual technological artifacts.

Sharing Local Tactical Advice and Technological Narratives

Champion's manual helped readers build a Locost (pronounced *low cost*) car, a project with reappropriation and making-do as its essence. As Champion stated emphatically, the Locost was "most definitely not" a kit car but a built-from-scratch replica of a Lotus Seven (p. 12). As described by one builder

> The premise of the book (and if you get into Locosting you will get to know it as "The Book") is that builders can inexpensively weld their own chassis … , panel it with aluminum, install the running gear from a donor car, and have themselves a passable replica of a Lotus 7. (Bura, 2003)

The result was a lightweight, minimalist sports car—with no roof, doors, windows, heat, or air conditioning, but with performance equal to expensive and exclusive factory-built sports cars such as Ferraris and Porsches.

As with the response to Muir's manual, this movement arose from local technological narratives, specifically Champion's own story. In his introduction, Champion recounted his son's frustration at being unable to afford a kit car. Champion recollected his experiences with his friends' Austin specials in the 1950s, noting that they took advantage of cheap wrecks, war-surplus materials, and tools to create their sports cars. Then Champion recounted suggesting to his son that they build a car from scratch themselves: " 'Why buy the kit? I bet we could make one for about £250.' And make one we did" (p. 10). They found a junked, rear-wheel-drive British Ford Escort in someone's yard, offered £25 for it, and found the remaining parts at junkyards and autojumbles (automotive flea markets). After a year of hard work and scrounging, they completed the car, which his son now races as an amateur. Champion told readers that his book derived directly from that experience, as well as his experience overseeing several students building Locosts at the school where he taught shop. Stories such as these lead into a manual that is less a set of traditional instructions than a structured discussion of tactics for building, primarily in the form of bricolage, or making-do—especially in the difficult matter of finding appropriate donors and parts.

Champion's stories centered on two interlocking themes: the freedom derived from tactical action and the capability of the user to produce his or her own products from the detritus of the strategic, industrial world. Champion emphasized the freedom that comes from this tactical approach: "I have yet to find two cars which have been made exactly alike. You have total freedom to adopt and adapt the basic plans to suit your circumstances and purposes" (p. 10). For example, because the chassis plans in Champion's manual made for a relatively narrow and short car, drivers weighing over 180 lbs. or standing over 5 ft. 10 in. had a hard time fitting into the cockpit. So builders typically have adapted Champion's frame plans to fit their own frames. In this way, Champion invited readers to inhabit his book and the car itself, making both texts their own.

Locost builders highly value this ability to write their own car. As builder Ken Walton put it (in an echo of de Certeau's example of the city traveler), there are "many paths to a common goal" (2000a): "The neat thing about the locost is that each one of us can build it his Way—room to be free!!! there are better ways and some not so good but let there be choice—that is where the fun is" (2000b). In Johnson's (1998) terms, these users have become builders and even designers, making their own tactical choices about how best to match technology to their needs and desires; they have started building and driving.

Encouraging User-Producer Documentation

Locost builders have produced not only cars, but also technical documentation. Beetle owners, as noted above, have documented their experiences with technology in the form of technological narratives or anecdotes. Locost builders have adopted the role of textual user–producer more fully, producing elaborate technological narratives and traditional technical documentation.

Typically, these documents (narrative or procedural) appear on *build sites*, websites where builders record their experiences and collect or produce documentation that might be useful to other builders. Build sites are remarkably varied but have grown into a genre with relatively consistent conventions. They often begin with a narrative of the inception of the build, laying out the builder's individual (and sometimes idiosyncratic) rationale and tactics for the project. They usually include a build diary, a blog chronicling the builder's efforts in the project. Some build diaries are visual—albums of captioned photographs requiring a high level of reader expertise to understand. But others are elaborate and revealing accounts of wrestling with challenges and coming up with tactical responses. Builders record not only their direct experience in building their cars, but also their tactical negotiations with the varying demands of spouses, children, employers, and the vagaries of life: bankruptcies, births, illnesses, and deaths (see, for example, *Doug's Locost*, 2004; Ogden, 2003). These technological narratives form an appropriation of Champion's ur-narrative, spreading that story in many individual permutations. Like Champion's narrative, their themes focus on the freedom of tactical action and the desire to learn how to produce a successful car that meets their own needs and desires.

As user-producers, Locost builders have also created a significant body of technical documentation, collected on build sites and online forums, such as "Locost," "Locost_North_America," and "Locost_Theory" (http://groups.yahoo.com). At an informal level, this documentation appears as lore and advice, such as one sees on any electronic mailing list. But recognizing the repetition of this kind of information, users have also developed more formal technical documentation, such as plans and instructions for making special tools, developing welding techniques, and registering the car. Others have offered highly technical advice, producing computer-aided-design (CAD) versions of the plans or documenting new techniques for suspension design (Bura, n.d.; McSorley, n.d.; Polan, 2004). An Australian builder (*An Analysis of Car Chassis*, n.d.) offered a report on his Finite Element Analysis (FEA) of the Locost chassis, including recommended structural modifications and improvements. Many of these formal documents are modeled after those often seen in technical communication: reports, specifications, procedures, and instructions. In essence, they form an appropriation of strategic practices to a tactical agenda. But attempts to strategize too formally can meet resistance: A *Locost Build Manual* on the Locost group files page (http://autos. group.yahoo.com/group/Locost_North_America/files) has the framework of a formal manual to which users can contribute their expertise, but it has never grown beyond a basic outline (Grossmith, 1999).

Contributing to Wider Cultural Narratives of Tactical Resistance

The products of the Locost movement—texts in the form of technical documentation, technological narratives, and the cars themselves—also reflect and contribute to wider cultural narratives of tactical resistance. One of these arose from the story of Colin Chapman, the designer of the Locost's inspiration, the Lotus

Seven. This car's story in itself is one of appropriation and tactics. In a pattern Champion echoed with the Locost, Chapman was known initially for building specials such as the Lotus Six, a cheap race car built by cannibalizing another car and attaching its parts to a stripped-down frame. The Lotus Seven, the follow-on to the Lotus Six, was a development of this tactical practice. Rather than offering the car fully built, Chapman produced the Lotus Seven as a kit because doing so allowed owners to avoid a British tax on prebuilt cars—a case of la perruque.

As a result of these tactics, the Lotus Seven was frequently seen as a symbol of freedom and rebellion from standards—in fact, many people first saw the Lotus Seven in the opening credits of the 1967–1968 British Broadcasting Corporation cult hit *The Prisoner* (Markstein, Tomblin, & Chaffey, 1967), which had similar themes. The show's protagonist, a recently resigned spy known only as Number 6 (note the allusion to the Lotus Six), is kidnapped and taken to the mysterious and technologically advanced Village, where he and others are held for knowing too much. His jailors constantly question Number 6 why he resigned, but he refuses to answer, crying out famously, "I will not be pushed, filed, stamped, indexed, briefed, debriefed, or numbered—my life is my own!" and "I am not a number; I am a free man!"

The Prisoner's (Markstein et al., 1967) cult status arose from Number 6's resistance to an all-seeing technocratic authority. A number of Locost builders have included in their build sites comments that *The Prisoner* has a close relationship to their concept of the car's meaning. One builder, for example, included a picture of the show's Lotus on his website, the caption reading

> I'm sure this is the true inspiration for my "Locost," … The Lotus 7 that appeared in the Late Sixties cult program "The Prisoner" …. I am not a number I am a free man!! In an age that did not know speed ramps or speed cameras, this TV series was about the "Individual" against the "System." I'm almost certain that EVERY "Locost" builder is an individual.!!!! (Osborne, n.d.)

Locost builders have extended this narrative for a pretechnological freedom into a postindustrial age, responding tactically to institutional forces that have imposed increasing controls on individual agency. Feenberg (1999) pointed out (from Bruno Latour) that "technical devices embody norms that serve to enforce obligations"—for example, a mechanical door closer takes over the responsibility of closing the door for humans who walk through it (p. 85). In this regard, Locost builders see their cars—and the technical narratives they tell about building them—as a tactical resistance against authoritarian control. Rather than driving cars that enforce conformity and remove direct agency (through cruise control, speed limiters, power steering, etc.), Locost builders want to build and drive cars embodying direct agency with technology.

DOCUMENTING RESISTANCE

The relationship to technology embodied by the Locost community, its technical narratives and documents, and its artifacts centers on a tactical resistance to strategic authority, much in line with de Certeau's (1984) description of everyday tactical practices in strategically determined systems. Working from Johnson's (1998) concept of user-as-producer, one can see that this resistance has become a greater possibility with the Locost than with the Beetle primarily because of the permeability between user and producer that the Web allows. Bolter (2003) commented that the Web allows people to enact resistance by creating their own publications, becoming producers, as well as consumers:

> The World Wide Web draws people into the production process on a much larger scale than television or film has ever done. ... As a consumer, one can only redirect the intended effects of media artifacts, but as a producer one can change the artifacts themselves. (pp. 23, 27)

Particularly in the Locost community, users become producers of documents and artifacts that subtly resist authority. This community depends on the ability to work around the strategic power of institutions and to scavenge, live, and drive through repairing and remaking what society has discarded. As extra-institutional systems, these communities spontaneously create and mutate technical documentation, thus taking their technological destiny into their own hands. One could even argue that in forming their own communities, Locosters rise above user-as-producer to the level of user-as-citizen—not by participating in the designs of strategic authority, but by using technology to form tactical communities with no place per se.

This dynamic is not without irony, especially in regard to the fantasies both the Beetle and Locost communities share: fantasies of escape from the power systems of industrial societies, of individual power within a context of cultural powerlessness ("I am not a number!"). In both cases, users buy into the human/technological narratives promoted by the manuals and are thus shaped by the larger cultural dynamics surrounding the manuals—an element brought to light by the fact that most who praise Muir's manual are liberals of a certain age and that most people reading Champion's manual are middle-class, middle-aged white men. By recognizing these dynamics, one can see both cultures centered on fantasies of freedom from the constraints of industrial society. Muir's manual reflected a desire to live cheaply and independently, freed from the constraints of the so-called establishment. Champion's manual revealed readers participating in a cultural narrative of an escape from middle-class limitations to enjoy upper-class luxuries like sports cars.

Examining technical documents and user communities such as these can reveal how people interact with technical documentation culturally—how deeply documentation can be integrated into the lives and fantasies of people in contemporary culture as they go beyond user-as-practitioner to user-as-producer and user-as-citizen. Observing such dynamics can provide some valuable insights about how technical documents convey and contribute to the forces of cultural power—whether centered around institutions, individuals, or dynamic communities.

IMPLICATIONS FOR TEACHING AND PRACTICE

These cases of extra-institutional, tactical technical communication offer interesting implications for work in technical communication pedagogy and research. In terms of technical communication pedagogy, we should continue efforts to broaden teaching so students can learn not only to participate in strategic workplace communication, but also to seize control of the possibilities for tactical technical communication, both within and outside of the workplace. We are already teaching students the technical skills they need to succeed as user-producers, but we should redouble our efforts to teach students how to think critically and independently about communication at a range of sites. As Bushnell (1999), Johnson-Eilola (2004), and others have pointed out, thinking of technical communication pedagogy primarily as workplace training undercuts our ability to foster savvy, flexible, tactical communicators. With the inevitable growth in outsourcing, training professional technical writers to fulfill limited, industrial roles will likely relegate many to underemployment, as Johnson-Eilola (2004) suggested: "If technical communicators do not take action to change their current situation, they will find their work increasingly contingent, devalued, outsourced, and automated" (p. 187). And as Savage (2004) argued, we should train technical communicators as sophists, "not mastering but negotiating continually shifting technologies, institutions, discourses, and cultures" (p. 189). This approach might seem to foster ethical uncertainty, but acknowledging the varied dynamics of power and the possibilities for tactical action in strategic power systems should provide further opportunities to discuss ethics with students—that is, ethics as a set of socially significant decisions they must make in negotiation with personal (tactical) and institutional (strategic) codes.

One way to encourage technical communication students to approach technical communication tactically is by asking them to look carefully at communication practices both inside and outside the workplace; as de Certeau (1984) pointed out, bricolage and la perruque have been going on for a long time in many settings. For example, introducing students to ethnographic scholarship—or as Henry (2000) did, actually asking students to engage in autoethnography—might encourage them to think about the relationship between the competing agendas and dynamics of tactical and strategic power.

The cases I have examined here also testify to the growing importance of technical communication in everyday life as a matter of production as well as consumption. In response, we teachers might renew our dedication to the technical communication service course, focusing on the many possible relationships (both strategic and tactical) between technology and culture (along with the basic technical writing skills such courses typically cover). Of course, we should continue to teach students about institutional communication practices, but with the goal of teaching how to work not only within but also through and at times around those networks of power. The industrial corporations of Henry Ford and Colin Chapman, to say nothing of the postindustrial corporations of Bill Gates and Steve Jobs, began with people tinkering in garages. We should enable all students with the technical communication skills to articulate their own contributions to and thoughts about technology.

In terms of research, we should continue to look at the cultural significance of technical documentation, particularly as it arises outside of organizations such as corporations and government agencies. Given our dependence on technology, technical documentation will inevitably happen in the shadows, its significance and role in culture yet to be exposed. Examining that extra-institutional documentation can help us understand not only our own field, but also the relationship between technology, discourse, and people's lives. In taking up this enlarged field of research, we should examine not just the safe cases, but the dangerous ones. Examples might include the documentation of computer-game cheating, hacking, tax evasion, and—given our historical moment—even terrorism manuals. These topics are not polite, but they help form the postindustrial world as much as the corporate and governmental technical documents that are typically the focus of technical communication research. Even in examining more traditional documents, we should continue to ask not just how to improve technical communication, but what technical communication and other technologies mean to users culturally, as well as how user-producers make further meaning through these cultural materials.

AUTHOR NOTE

I have used the 1969 edition of Muir's manual and the 2000 edition of Champion's manual throughout this article.

REFERENCES

Amazon.com. (n.d.). *Customer reviews.* Retrieved October 11, 2005, from http://www.amazon.com/gp/product/customer-reviews/1566913101

An analysis of car chassis. (n.d.). Retrieved October 11, 2005, from http://locost7.info/files/chassis/kitcaranalysis_V2.doc

Avalon Travel Publishing. (n.d.). *How to keep your Volkswagen alive.* Retrieved October 11, 2005, from http://www.travelmatters.com/vw/stories.html

Barton, B., & Barton, M. (1988). Narration in technical communication. *Journal of Business and Technical Communication, 2,* 36–48.

Blyler, N., & Perkins, J. M. (1999). *Narrative and professional communication.* Stamford, CT: Ablex.

Bolter, J. D. (2003). Theory and practice in new media studies. In G. Liestol, A. Morrison, & T. Rasmussen (Eds.), *Digital media revisited: Theoretical and conceptual innovation in digital domains* (pp. 15–33). Cambridge, MA: MIT Press.

Bura, P. (2003, May 2). *Pete's rotary Locost.* Retrieved October 11, 2005, from http://www.geocities.com/phbura@ameritech.net/index.html

Bura, P. (n.d.). *Designing a double wishbone suspension: A 3-position graphical approach.* Retrieved October 11, 2005, from http://www.geocities.com/phbura@ameritech.net/geometry.html

Bushnell, J. (1999). A contrary view of the technical writing classroom: Notes toward future discussion. *Technical Communication Quarterly, 8,* 175–188.

Carpenter, H., & Dombrowski, P. (1997). *Ethics in technical communication.* Boston: Allyn & Bacon.

Champion, R. (2000). *Build your own sports car for as little as £250* (2nd ed.). Somerset, England: Haynes.

de Certeau, M. (1984). *The practice of everyday life* (S. Rendall, Trans.). Berkeley: University of California Press.

Doug's Locost. (2004). Retrieved April 15, 2005, from http://www.geocities.com/doug1399

Dubinsky, J. (Ed.). (2004). Civic engagement [Special issue]. *Technical Communication Quarterly, 13*(3).

Feenberg, A. (1999). *Questioning technology.* New York: Routledge & Kegan Paul.

Grabill, J. T. (2003). Community computing and citizen productivity. *Computers and Composition, 20,* 131–150.

Grossmith, D. (1999, May 26). *Locost build manual.* Retrieved October 11, 2005, from http://finance.groups.yahoo.com/group/locost/files

Harrison, T., Zappen, J., & Prell, C. (2002). Transforming new communication technologies into community media. In N. W. Jankowski and O. Prehm (Eds.), *Community media in the information age: Perspectives and prospects* (pp. 249–269). Cresskill, NJ: Hampton.

Harrison, T. M., & Katz, S. M. (1997). On taking organizations seriously: Organizations as social contexts for technical communication. In K. Staples and C. Ornatowski (Eds.), *Foundations for teaching technical communication* (pp. 18–29). Greenwich, CT: Ablex.

Henry, J. (2000). *Writing workplace cultures: An archaeology of professional writing.* Carbondale: Southern Illinois University Press.

Houp, K. W., Pearson, T. E., Tebeaux, E., & Dragga, S. (2002). *Reporting technical information* (10th ed.). New York: Oxford University Press.

Johnson, R. R. (1998). *User-centered technology: A rhetorical theory for computers and other mundane artifacts.* Albany: State University of New York Press.

Johnson-Eilola, J. (2004). Relocating the value of work: Technical communication in a post-industrial age. In J. Johnson-Eilola & S. Selber (Eds.), *Central works in technical communication* (pp. 175–192). Oxford, England: Oxford University Press.

Johnson-Eilola, J. (2005). *Datacloud: Toward a new theory of online work.* Retrieved October 11, 2005, from http://people.clarkson.edu/~johndan/read/datacloud/index.html

Katz, S. B. (1992). The ethic of expediency: Classical rhetoric, technology, and the holocaust. *College English, 54,* 255–275.

Longo, B. (1998). An approach for applying cultural study theory to technical writing research. *Technical Communication Quarterly, 7,* 53–73.

Longo, B. (2000). *Spurious coin: A history of science, management, and technical writing.* Albany: State University of New York Press.

Make: Technology on Your Time. (2005, March). *1.*

Markel, M. (2004). *Technical communication* (7th ed.). Boston: Bedford/St. Martin's.

Markstein, M., & Tomblin, D. (Writers), & Chaffey, D. (Director). (1967). Arrival [Television series episode]. In P. McGoohan (Producer), *The Prisoner.* Penrhyndeudraeth, Wales: Independent Television Commission.

McSorley, J. (n.d.). *CAD drawings.* Retrieved October 11, 2005, from http://www.mcsorley.net/locost

Meyer, P. R., & Bernhardt, S. A. (1997). Workplace realities and the technical communication curriculum: A call for change. In K. Staples & C. Ornatowski (Eds.), *Foundations for teaching technical communication* (pp. 85–98). Greenwich, CT: Ablex.

Muir, J. (1969). *How to keep your Volkswagen alive! A manual of step by step procedures for the compleat idiot.* Santa Fe, NM: John Muir Publications.

Muir, J. (1973). *The velvet monkey wrench.* Santa Fe, NM: John Muir Publications.

Nelson, W. H. (1965). *Small wonder: The amazing story of the Volkswagen.* Boston: Little, Brown.

Ogden, P. (2003). *Locost construction diary.* Retrieved October 11, 2005, from http://locost7.info/diary/diary1.php

Ornatowski, C., & Bekins, L. K. (2004). What's civic about technical communication? Technical communication and the rhetoric of "community." *Technical Communication Quarterly, 13,* 251–269.

Osborne, D. (n.d.). *Inspiration. Locost build diary.* Retrieved October 11 2005, from http://www.groovy42.freeserve.co.uk/what.htm

Polan, M. (2004, July 21). *Using a CAD program to analyze your suspension.* Retrieved October 11, 2005, from http://www.7builder.com/SuspensionGeometry/SuspensionDynamics.htm

Rosen, M. J. (Ed.). (1999). *My bug.* New York: Artisan.

Rude, C. (Ed.). (2000). The discourse of public policy [Special issue]. *Technical Communication Quarterly, 9*(1).

Russell, D. R. (1997). Rethinking genre in school and society: An activity theory analysis. *Written Communication, 14,* 504–554.

Savage, G. J. (2004). Tricksters, fools, and sophists: Technical communication as postmodern rhetoric. In T. Kynell Hunt & G. J. Savage (Eds.), *Power and legitimacy in technical communication. Strategies for Professional Status* (Vol. 2, pp. 167–193). Amityville, NY: Baywood.

Scott, J. B. (2004). Rearticulating civic engagement through cultural studies and service-learning. *Technical Communication Quarterly, 13,* 289–306.

Subbiah, M. (1997). Social construction theory and technical communication. In K. Staples & C. Ornatowski (Eds.), *Foundations for teaching technical communication: Theory, practice, and program design* (pp. 53–65). ATTW Contemporary Studies in Technical Communication Series, Vol. 1. Greenwich, CT: Ablex.

Walton, K. (2000a, 17 October). Donor [202]. Retrieved October 11, 2005. Message posted to http://autos.groups.yahoo.com/group/Locost_North_America/messages

Walton, K. (2000b, 9 October). Stuff [149]. Retrieved October 11, 2005. Message posted to http://autos.groups.yahoo.com/group/Locost_North_America/messages

Miles A. Kimball is an assistant professor in Texas Tech University's Technical Communication and Rhetoric Program. He is the author of *The Web Portfolio Guide* (Longman 2003) and is writing a textbook in document design (Bedford/St. Martin's Press) and a book on the history of information graphics.

Rhetorical Agency, Resistance, and the Disciplinary Rhetorics of Breastfeeding

Amy Koerber
Texas Tech University

Drawing on interviews from a qualitative study, this article extends theorizing about rhetorical agency and resistance by analyzing how breastfeeding advocates and their clients resist medical regulatory rhetoric. The resistant acts that interviewees describe begin with a negotiation of discursive alternatives and subject positions framed by the grid of disciplinary rhetoric about breastfeeding. But in some acts of resistance, breastfeeding women use both discursive and bodily actions to disrupt the intelligibility of this grid and what it deems possible. When such disruption occurs, the results are unpredictable and so must be understood as more than the occupation of preexisting subject positions.

You're challenging the person that's taking care of your baby ... it's almost like when I told the dental hygienist that I wanted her to change her gloves ... the thought that she had a sharp instrument went through my head ... do I really want to say this?

Katherine, breastfeeding advocate
(personal communication, May 8, 2000)

Technical communication scholars who incorporate theories and methods from cultural studies have asserted that an awareness of the capacity for rhetorical agency, resistance, and change distinguishes research in the technical communication field from similar projects in cultural studies (Herndl, 2004, p. 6; Reeves, 1996, p. 131; Scott, 2003, p. 33). Despite these assertions, however, there remains some uncertainty about the precise nature and limitations of rhetorical agency and resistance. Although it is well established that subjects are capable of "rhetorical negotiation" (Reeves, p. 153) or "choices between competing alternatives" (Britt, 2001, p. 147), it is also clear that this ability to choose has limits. Furthermore, it is clear that such agency cannot be reduced to subject-centered, strategic use of language directed against ideological force in a two-way struggle, but rather must be understood as partial and as closely implicated with the same discursive structures that embody such force (Britt; Herndl; Reeves; Schryer, Lingard, Spafford, & Garwood, 2003; Scott).

Taking these tensions surrounding rhetorical agency and resistance as a starting place, this article investigates breastfeeding discourse as "disciplinary rhetorics," that is, as "discursive bodies of persuasion that work with extrarhetorical actors to shape subjects and to work on and through bodies" (Scott, 2003, p. 7). The concept of disciplinary rhetoric derives from Foucault's (1977) concept of disciplinary power, a uniquely modern form of power that emerged in Western civilization during the 18th century as part of an array of political, institutional, and cultural shifts. Although disciplinary power, as Foucault described it, is less visible than the more obvious forms of power that preceded it (corporal punishment, for instance), its ultimate effect can be to heighten the visibility of subjects' bodies in relation to institutions such as medicine.

My analysis is based on interviews with breastfeeding advocates who work in various capacities to support breastfeeding mothers and assist them when they encounter problems. Interviewees suggest that to follow current official medical guidelines on breastfeeding, which advise that infants should be breastfed for at least 1 year (American Academy of Pediatrics [AAP], 1997), a woman has to resist various other elements of medical discourse and larger cultural perceptions that directly contradict these official guidelines. As the remarks of the women interviewed suggest, the disciplinary power of breastfeeding discourse lies not in any single message about what women should do, but in the mixed messages that circumscribe what their bodies can do.

Echoing the findings of previous technical communication research, my study suggests that when we researchers adopt the critical stance that cultural studies approaches entail, we must account for rhetorical agency without reducing such agency either to the occupation of preexisting subject positions or to strategic, subject-centered language use that enables transcendence of such predefined positions. Although we might feel compelled to choose between privileging individual agency or ideological force, my analysis reinforces the idea that the two are inextricably linked and adds to our understanding by taking a close look at the relationship between the rhetorical agency involved in acts of resistance and the ultimate outcomes of such acts. Specifically, I argue, the acts of resistance that interviewees describe begin as active selection among discursive alternatives—the kind of rhetorical negotiation that might be construed as the occupation of preexisting subject positions rather than true resistance. However, when we examine the effects of these women's acts of resistance, we see something more than the occupation of preexisting subject positions; we see that resistance can "defy translation, throw sense off track, and, thus, short-circuit the system through which sense is made" (Biesecker, 1992, p. 357).

After a brief discussion of my interview methods and the study's background, I examine how breastfeeding discourse operates as a disciplinary rhetoric. I then investigate how the rhetorical agency and acts of resistance that interviewees discuss can be accounted for in theoretical terms.

INTERVIEWS AND STUDY BACKGROUND

The interview excerpts presented in the article come from a larger study of the 1997 AAP policy statement, "Breastfeeding and the Use of Human Milk." This statement attained a high profile in the medical community and in the mass media because it urges women to breastfeed for at least the first year of the infant's life, but further because it stipulates that breastfeeding is "the reference or *normative* model against which all alternative feeding methods must be measured with regard to growth, health, development, and all other short- and long-term outcomes" (p. 1035). In the larger study from which this article derives (Koerber, 2002), I examined a range of medical texts published between 1940 and the present, including research articles, advice literature, and policy documents, and I interviewed 17 people involved in breastfeeding advocacy to better understand both the exigency that led to the statement's declaration of breastfeeding as the norm and the rhetorical effects of this declaration. The quotations included in this article are drawn from the subset of data in which interviewees explicitly discussed the resistance they enact and/or help other mothers enact against the rhetorical constructions of breastfeeding that circulate in medical discourse and larger cultural perceptions, many of which continue to treat bottle-feeding as the norm in spite of the most recent AAP recommendations. The people interviewed all consider themselves breastfeeding advocates, although they practice their advocacy in different locations—some within the medical community, and some beyond.

The interviews were conducted between May and August 2000. In this article, I use pseudonyms to protect all interviewees' identities. Each interview lasted between 1 and 3 hours.[1] Because this is a qualitative study, neither the larger body of data nor the subset that this article presents is intended to statistically represent any particular demographic group. Nonetheless, it is important to acknowledge that the interview sample is much more representative of women who have had positive breastfeeding experiences than of any other group; most of the women I inter-

---

[1] This study was approved by University of Minnesota's Institutional Review Board and followed all policies and procedures applicable to human subjects research. When conducting the interviews, I made audio recordings of all but one of the interviews and then transcribed the tapes myself. The exception was one interviewee who did not wish to be recorded. For this interview, I took detailed notes and attempted to code these notes according to the same process I used to code the interview transcripts.

viewed had successfully breastfed babies before they became breastfeeding advocates, and even within this group, there is a bias toward women who were enthusiastic enough about breastfeeding to become involved in advocacy efforts. I admit, in the present study, that I have made no effort to consider the perspectives of women who never attempted breastfeeding or had less positive experiences with it. Such consideration would undoubtedly reveal many other ways in which women resist the disciplinary power of breastfeeding discourse, and it would shed new light on the difficulties inherent in such resistance. However, it is beyond the scope of this article's analysis.

BREASTFEEDING DISCOURSE
AS DISCIPLINARY RHETORIC

The AAP's 1997 declaration of breastfeeding as the norm was an effort to make infant-feeding information more consistent within the medical community and in the culture at large. As recent historical research has revealed, in the early and mid 20th century, U.S. hospitals adopted practices such as separating mothers and infants, enforcing rigid feeding schedules, and offering formula at the first sign of breastfeeding difficulty (Van Esterik, 1995, p. 147). Because a mother's milk supply can be established and maintained only through frequent infant sucking, practices such as these can interfere with breastfeeding by reducing the mother's chances of establishing a healthy milk supply before she even leaves the hospital (Quandt, 1995, p. 134–135). Thus, many scholars believe breastfeeding mothers have long faced a contradiction: Although official medical discourse has long promoted breastfeeding, bottle-feeding has become accepted as the medical and cultural norm (Apple, 1987; Blum, 1999; Hausman, 2003; Maher, 1992; Yalom, 1997).

In response to this historical context, the 1997 AAP statement offered specific advice to govern physicians' and mothers' interpretations of the breastfeeding body—advice intended to reassure them in the face of infant behavior that might seem worrisome in a culture accustomed to bottle-fed babies. For instance, the statement affirmed that it is normal for breastfed babies to eat frequently, advising mothers to allow their babies to eat "whenever they show signs of hunger" (AAP, 1997, p. 1036), rather than try to adhere to a rigid feeding schedule, and it spelled out what health care practitioners should expect to see in the normal nursing mother–baby pair. Although this policy statement has achieved a high profile in the medical community and in the popular media, according to the breastfeeding advocates I interviewed, women who attempt to breastfeed today still face the long-standing rhetorical construct of bottle-feeding as the norm, disguised in medical and cultural practice as neutral scientific truth. In fact, my study suggests, disciplinary power does not reside exclusively in the language of official medical discourse such as the AAP's 1997 statement, but in the convergence of various, often conflicting, messages that mothers still receive from the medical establishment and the culture at large.

In theoretical terms, the conflicts that interviewees described manifest Foucault's (1977) observations about the connection between power and knowledge. Foucault urged researchers to acknowledge that "power and knowledge directly imply one another; that there is no power relation without the correlative constitution of a field of knowledge, nor any knowledge that does not presuppose and constitute at the same time power relations" (p. 27). Although Foucault's concept *pouvoir-savoir* is usually translated as *power-knowledge*, *pouvoir* means *to be able* as well as *power* (Spivak, 1992, cited in Biesecker, 1992, pp. 355–356). When it is translated as *to be able*, "power names not the imposition of a limit that constrains human thought and action but a being-able that is made possible by a grid of intelligibility" (Biesecker, p. 356). Adopting this slightly different translation, disciplinary power can be understood as not only dictating what subjects should do, but also as producing the very rhetorical situations in which they act by specifying what their bodies can do.

Reflecting this idea that disciplinary rhetoric can define what subjects' bodies are capable of doing, some of my interviewees suggested that the conflicting messages women receive sometimes make breastfeeding impossible. For instance, Sharon, a lactation consultant in a large, urban hospital, said that, in the hospital where she works, breastfeeding is sometimes still "sabotaged." That is, even though the AAP's 1997 policy statement stipulated that most mothers can nourish their babies exclusively on breast milk for the first 4 to 6 months of life and advises physicians that supplementary foods should not be given to newborns "unless a medical indication exists" (p. 1036), nurses sometimes offer bottles of formula anyway. "There's lots of ways that [breastfeeding] can be sabotaged. And certainly I don't want to give you the wrong impression that it's every nurse who's doing this … but it is a problem" (Sharon, personal communication, March 31, 2000). Despite the critical tone of Sharon's remarks, health care practitioners who offer unnecessary supplements do not necessarily act with the intention of disrupting the breastfeeding relationship. More likely, they offer bottles because it is convenient and it accords with past practices. But regardless of the motivations for it, my interviewees saw early supplementation, especially if it occurs without the mother knowing about it, as particularly insidious because the breastfeeding relationship is especially vulnerable to disruption in the first few days of an infant's life.

This potential for problematic advice in the hospital environment is compounded by the fact that women often attempt breastfeeding for the first time with few mental pictures of others who have breastfed babies and, therefore, with no point of reference to indicate what breastfeeding is supposed to look like and how the breastfed baby is supposed to act. This lack of a cultural reference point converges with the conflicting messages women sometimes receive in medical settings, further circumscribing what the breastfeeding body can do and making it even more difficult to follow the AAP's recommendation to breastfeed for at least 1 year.

As Sharon observed, the long-standing tendency for women to see far more examples of bottle-feeding mothers than breastfeeding mothers often leads breastfeeding women to doubt the capabilities of their own bodies. "Most people have enough milk, ... [but] most people think they don't have enough milk ... so perceived insufficient milk is a big problem" (personal communication, March 31, 2000). She elaborated on the subject, claiming that women's doubts about their bodies can ultimately interfere with lactation itself: "It's a leap of faith to think that your body can nourish a baby ... oftentimes, the mom's insecurities will lead to supplementation and ultimate weaning" (personal communication, March 31, 2000).

As Diane, a postpartum doula,[2] said, pregnant women often take a breastfeeding class offered by the hospital where they plan to give birth. However, this in itself is not enough to make breastfeeding seem like a normal activity because many women "grew up without seeing enough women breastfeeding, so we're not internalizing the images of breastfeeding" (personal communication, April 24, 2000). Michelle, a La Leche League leader, echoed Diane, asserting that our "culture has set a lot of women up [to fail] at breastfeeding" because expectations that infants should adopt regular sleep patterns and admonitions against "pacifying the baby with the breast" are based on behaviors more typical of bottle-fed than breastfed babies (personal communication, June 1, 2000).

The situations these women described can lead to serious breastfeeding problems, and whether these problems are imagined or real, they can lead the mother to perceive it as necessary to wean from breast to bottle, even if weaning is not medically warranted, and even if she does not want to do it. In theoretical terms, then, breastfeeding advocates' remarks about the ways that health care practitioners, or even friends and relatives, can undermine breastfeeding success offer an example of disciplinary rhetoric circumscribing what a woman's body is able to achieve. Even if women are told that they should breastfeed for at least a year, the institutional practice of offering unnecessary supplements, as well as cultural perceptions based on the model of bottle-fed babies, can make it difficult or impossible to do so. As their remarks indicate, even if women today are educated in breastfeeding techniques and the benefits of breastfeeding, they often receive

---

[2]Doulas are women hired to support mothers either during labor or during the postpartum period, or both. Although insurance generally does not cover their cost, doulas are increasingly used by women who can afford to pay for this service out of pocket. Although many doulas are self-trained, a training program sponsored by Doulas of North America offers certification (see the organization's website at http://www.dona.com).

conflicting advice from either medical authorities or friends and relatives that undermines successful breastfeeding, literally making their bodies unable to lactate successfully.

"BUCKING THE SYSTEM"

As suggested in the quotation with which this article opens, resisting disciplinary rhetoric, especially when it involves direct questioning of a medical authority, is difficult. Despite this difficulty, interviewees offered numerous examples to affirm that they and the mothers they assist sometimes resist the disciplinary rhetorics discussed in the previous section. In the words of one La Leche League leader, to breastfeed a baby in U.S. society a woman has to "buck the system." Specifically, the women in my study reported instances in which they exercised some degree of choice among the competing discourses that form the framework of meaning established by the disciplinary rhetorics of breastfeeding. For instance, Sharon, a lactation consultant in a large urban hospital, recounted how she has used the 1997 AAP policy statement to educate the physicians in her hospital and even to bolster her credibility in disagreements with such physicians:

> I have made stacks of copies [of the AAP 1997 statement] and put them in the resident lounge. I have slipped one under the door of our chief perinatalogist when he differed with me on a certain point …. There was an instance where I was telling a mom that, for the most part, breastmilk is all the baby needs for the first 6 months, and there was a doctor behind the curtain, who then came around the curtain, and he said, "But that may not be true for all babies; check with the baby's doctor on that," and I gave him a copy [of the 1997 statement]. (personal communication, March 31, 2000)

In contrast to Sharon's reliance on the AAP (1997) policy statement to challenge physicians' authority, some interviewees emphasized the unique perspective that the distinctly nonmedical knowledge of an organization such as La Leche League provides to breastfeeding women. In fact, many interviewees claim that their own needs for alternatives to mainstream medical discourse when they began breastfeeding led them to become La Leche League leaders and/or lactation consultants in the first place.

For instance, according to Janet, a board-certified lactation consultant and La Leche League leader, she first attended La Leche League when her first daughter was young because it provided an alternative understanding of the breastfeeding relationship, allowing her to reject the emphasis on scheduled feedings that she had learned through mainstream medical discourse:

I enjoyed the information and support that I got. Among other things, I needed to hear that I didn't have to let babies cry in order to wait a certain number of hours between feedings, so we had a lot less crying at our house when I started going to meetings, and I started to trust myself a little more as a mother instead of just coming more from my nursing background of doing things on a schedule. (personal communication, April 19, 2000)

Similarly, Kim, a La Leche League leader, described the role of La Leche League as filling in the gaps in mainstream medical practice, providing the "practical stuff that women need, the actual nuts and bolts of how the relationship works and what to do when there's problems" (personal communication, June 13, 2000).

Providing a more specific example, Jackie, also a La Leche League leader, told the story of one woman who had consulted her regarding a physician's recommendation for treatment of a thyroid problem. According to Jackie, this woman eventually opted for the treatment that preserved the breastfeeding relationship, even though this required questioning the physician's initial recommendations:

I know one woman who was told she needed to do the radioactive treatment to kill off her thyroid gland because she was hyperthyroid, and to do that ... she would have to wean her baby for ... a week or 10 days. And she said, "I don't want to do that. What else can we do?" And the doctor said, "Oh, well, this is the only way to treat the problem other than surgical removal of your thyroid gland." ... She said, "When I look at my options, this is a minor surgical procedure that's going to have a 2-hour separation between me and my baby versus something that's going to require me to wean for ... 14 days." She opted for the surgical removal of her thyroid rather than killing it off with a radioactive drug. (personal communication, July 11, 2000).

In this case, it was the willingness of a woman to directly question a physician's recommendations that allowed her to maintain breastfeeding. Along these lines, many interviewees noted that their job is not only to provide alternative information, but also to empower women so that they will feel confident enough to ask the kinds of questions that will solicit such information from their physicians. As Sharon says, an important part of her job is "to give [mothers] the confidence that they can provide enough milk for the baby, and be assertive for themselves when dealing with health professionals" (personal communication, March 31, 2000).

These accounts illustrate how rhetorical agency, defined as negotiation among competing alternative discourses, grants individuals some ability to reject discursive elements that they find problematic. This ability of individuals to enact some degree of choice among competing discourses has been noted by previous technical communication scholars (Britt, 2001; Reeves, 1996; Scott, 2003). But such scholars also note that this type of rhetorical agency does not necessarily allow subjects to escape the ideological force of institutional discourses. For instance, in her study of a 1987 Massachusetts mandate requiring insurance companies to cover infertility treatments, Britt (2001) invoked the concept of the double bind to explain "both how discourses create spaces within which individuals act and how individuals act within these spaces" (p. 13). Britt concluded that rhetorical agency led some women in her study to feel "empowered by working themselves out of double binds" (p. 147). But she ultimately adopted a cautious stance toward such agency and empowerment, noting that choosing among the alternatives they appear to offer does not necessarily resolve double binds and that some double binds might be entrenched in U.S. culture to such an extent that they are nearly impossible for individuals to resolve (p. 148). In his study of the discourses of HIV testing, Scott (2003) echoed Britt's observations, asserting that the individual examined in his study is "not a docile body straightforwardly shaped by the testing program's disciplinary rhetorics … [but is] conditioned by the national pedagogy, interpersonal relationships, and cultural norms" (p. 157). Scott concluded that subjects of HIV-testing discourse might have some ability to "negotiate the alternative notions of risk" conveyed in such discourse, but that it is hard to know how much negotiation will be possible in any given situation (p. 157).

These scholars' observations raise important questions about how to interpret the interviewees' accounts of resistance I have presented so far. In these accounts, it is clear that the interviewees' rhetorical agency cannot be understood simply as strategic, subject-centered language use that enables them to escape the grid of meaning established by disciplinary rhetoric. But if one acknowledges the extent to which this grid of meaning continues to circumscribe their actions, it is not clear whether the actions interviewees described truly count as resistance. Whether they appealed to alternative discourses within medicine, such as the AAP (1997) policy statement, to nonmedical discourses, such as La Leche League, or even to friends and family members, it could be argued that the alternative discourses these women marshalled are simply alternative forms of disciplinary rhetoric. Extending this line of reasoning, one could argue that the women are not really subverting disciplinary power but rather are choosing among several preexisting subject positions made available to them within its framework.

In short, it seems, to understand these women's acts of resistance, we need to see evidence that they can disrupt disciplinary power even without escaping its grip. This understanding of resistance echoes Foucault's (1978) assertion that even though resistant acts "can only exist in the strategic field of power relations," such acts do not have to be understood as just "a reaction or rebound, forming with respect to the basic domination an underside that is in the end always passive, doomed to perpetual defeat" (p. 95). In the accounts of resistance I have presented so far, it is not clear how this type of disruption might occur. The accounts of resistance I present next, however, provide clearer insights into such a possibility. In these next accounts, the women I interviewed discussed more specifically the relationships between their rhetorical agency and the effects of the resistance that this agency allows. Specifically, in these discussions, we begin to see that resistance might initially involve a form of rhetorical agency in which subjects simply occupy preexisting subject positions, but the effects of this agency—the acts of resistance—can disrupt the sense established by disciplinary rhetoric, exceeding the boundaries of these subject positions in unpredictable ways.

Some of the most striking examples offered by interviewees in this regard are those describing instances in which they shocked medical authorities, friends, or acquaintances by defying preconceived notions of what women's bodies are capable of doing. For instance, Karen, a lactation consultant in a large suburban hospital, talked about nursing twins several years ago and repeatedly having to address the astonishment of health care professionals and others who did not believe this should be possible:

> I have twins, and of course, everyone said, "Well, you're not gonna breastfeed twins, are you?" And so I had a really nice lady, a lactation consultant ... . She said, "You know what, tell them you have two breasts just in case you have two babies." And, so, I've hung onto that ever since. (personal communication, June 2, 2000)

As another example, Ellen, a La Leche League leader, talked about her earlier experience of breastfeeding a premature baby. Because premature babies are typically not able to breastfeed directly, a mother has to pump breast milk and feed it to the baby through a feeding tube. As Ellen's remarks make clear, this is not something that many mothers have done, and part of what made it possible for her was that she had the support of a lactation consultant in the hospital where she and her baby were:

> In looking back on it now, what I did with a preemie I guess is pretty rare. There aren't very many preemies who successfully breastfeed ... but, you know, I never got that perception when I was doing it. I was just pumping, and you know, and I just kept pumping. I was like, "What else am I gonna do?" ... I had nothing to do except get this baby home, and that was my goal. And, you know, even, once in a while the doctor would come in and say, "I'm very impressed that you're still keeping up with this. Most moms give up by now." ... And the lactation consultant there was very helpful, and the nurses were very helpful and encouraging. And so, I think that made a world of difference knowing that there was that support system there. (personal communication, July 5, 2000)

As these accounts illustrate, interviewees still depended on the kind of rhetorical negotiation among competing alternatives within the grid of meaning established by disciplinary rhetoric that could be construed as the occupation of preexisting subject positions. Specifically, in both accounts, the women emphasized that they needed the support of a La Leche League leader or lactation consultant to enable them to carry out the actions they considered resistant. But because these accounts refer specifically to the reactions of other people implicated in the grid of meaning established by disciplinary rhetoric, we also begin to see how this grid might be disrupted when subjects perform acts that this grid had previously defined as impossible. Recalling that disciplinary rhetorics define what subjects can do as well as what they should do, these women can be seen as disrupting sense by accomplishing things that disciplinary rhetorics dictated they could not do. The surprised reactions of bystanders, including medical authorities, friends, and relatives, can be read as an indication of the potential for such disruption.

Although it confronts a different set of issues than I have addressed so far, the issue of nursing in public places offers another useful example to illustrate the nuances of this relationship between rhetorical agency and resistance. More than one interviewee explicitly mentioned that a La Leche League meeting can be a good place to prepare a woman for her first experience with public nursing. As Diane, a postpartum doula, said,

> It's kind of a culture shock when you walk into a La Leche League meeting and see all these breastfeeding mothers, and you've never seen even one in your life, and all of a sudden you've got a room full of them, you know, one nursing a baby and a toddler on the other side. (personal communication, April 24, 2000)

Lisa, a La Leche League leader, explained, however, that despite this preparation, a woman is still on her own the first time she nurses in public outside of a league meeting:

> [A La Leche meeting] is a good place to go over that particular wall. And it is … like a wall when you first breastfeed in public, you know. You think all eyes are on you, you feel like you're 13, and so that can be a good place to come. (personal communication, May 22, 2000)

As the example of public nursing illustrates, resistance is not just a question of an individual selecting bits and pieces from various competing discourses. By consulting La Leche League, which resists mainstream medical discourse as well as broadly accepted social and cultural norms, a woman is empowered to resist the cultural norm that forbids public breastfeeding, but the ultimate effect of her choice is not scripted in the predefined subject position she selects. Specifically, at the moment she actually breastfeeds in public by herself for the first time, a woman is forced to act in a space where the counter-discourse of La Leche League does not prevail. It is at this moment, I would argue, she has the potential to disrupt meaning and sense.

This type of disruption can be further explained with reference to de Certeau's (1984) assertion that "the users of social codes turn them into metaphors and ellipses of their own quests" (p. xxi). De Certeau compared the individual use of social codes to "the rules of meter and rhyme for poets of earlier times: a body of constraints stimulating new discoveries, a set of rules with which improvisation plays" (p. xxi). In other words, the impetus to resist might start with an action that could be construed as simply occupying a preexisting subject position (or use of a preexisting "social code," in de Certeau's terms): identifying oneself as aligned with the cultural norms of La Leche League, rather than mainstream cultural norms, which do not widely condone public nursing. However, when one considers the moment at which a woman first attempts nursing in a public space outside of a La Leche League meeting, the occupation of preexisting subject positions is no longer an adequate account of her actions. The ultimate effect of asserting rhetorical agency in this type of situation is something more than the sense—or meaning—made possible by disciplinary rhetoric's grid of meaning because when a mother breastfeeds in public for the first time, she cannot know the effect of overcoming the wall. Whereas she knows she will be safe violating this cultural norm in the safe space of a La Leche League meeting, she does not know what she will face when she breastfeeds for the first time in a shopping mall, restaurant, or other public space.[3]

Thus in various ways, the forms of resistance described in this section involve rhetorical agency, but they also involve a moment when the subject does more than just select among competing alternatives within the disciplinary grid—a moment when the subject suggests that her actions disrupt the sense of this grid, or at least have the potential to do so. To further understand this potential for disruption and how it might contribute to the long-term effects of resistance, it helps to think beyond the isolated moment or event that each woman described. Scott (2003) called for this type of expansion in his stipulation that the "process of subject formation … cannot be located in a discrete event or set of events, and cannot simply be viewed as the product of that event's discourse or interaction between rhetorical agents" (p. 155). In other words, the rhetorical agency and acts of resistance described by the breastfeeding advocates in my study do not necessarily have to be seen as individual acts with distinct beginning and end points, but rather, might be understood as, at least potentially, having implications that extend beyond the individual events or acts they describe.

Although the present study is not equipped to systematically track the long-term effects of resistant acts in precisely the manner that this expanded view would require, the potential for long-term disruption of sense is at least suggested in one interviewee's observation that as infant-feeding norms gradually change, the feats

---

[3]On November 10, 2003, a Burger King employee at an Orem, Utah, restaurant asked a customer who was breastfeeding her baby to leave. The employee's request was reportedly a response to the remarks of another customer who claimed to feel uncomfortable with the mother's public breastfeeding. Burger King later issued an official apology (see "Burger King apologizes," 2003, for the full story). Although many states have laws protecting women's rights to breastfeed in public, incidents such as this are not at all uncommon.

women accomplish might come to be seen as less remarkable. Specifically, Jackie, a La Leche League leader, described an instance in which she challenged medical perceptions about what her body should be capable of doing:

> When my second one was born 10 and a half years ago, one of the nurses asked me, right after he was born—I had him latched on—she said, "Wow, you're a pro at this; when was the last time you nursed?" And I glanced at the clock and said, "Oh, about twelve hours ago." And that absolutely stunned them, that I had nursed through a pregnancy … . That was quite an outrageous thing for them to be confronted with because I know I wasn't the first woman to do that, but I was the first one to admit it. (personal communication, July 11, 2000)

In this description of an instance in which she defied medical perceptions of what is possible, Jackie's account resembles those of other interviewees. But Jackie also went on to explain that, in her opinion, this same act is now less likely to be seen as remarkable because more health care practitioners have encountered something like it:

> And now you find more nurses who know it's OK to nurse through a pregnancy, or they'll talk about, "Yeah, there was a mom in here. And she was nursing a newborn on one side and her 18-month-old nursing on the other side at the same time, and wasn't that sweet," rather than, "I can't believe the doctor allowed that." (personal communication, July 11, 2000)

Jackie's remarks suggest how the individual resistant acts that other interviewees described might over time have a profound impact in the lines of sense established by disciplinary rhetoric. Although Jackie's remarks pertain to an encounter with medical authorities, it is easy to imagine that they might apply to the issue of public breastfeeding as well; that is, we might speculate that if enough women continue to disrupt sense by violating cultural norms that forbid public breastfeeding, these norms will eventually change.

IMPLICATIONS FOR TECHNICAL COMMUNICATION AND CULTURAL STUDIES

Reinforcing the findings of previous technical communication research, this study suggests that the multiplicity of breastfeeding discourse makes resistance possible because it allows women some ability to construct their breastfeeding experiences through negotiation among its competing messages. In affirming this observation of previous research, I have also highlighted a dilemma that we researchers face as we integrate cultural studies theories and methods into technical communication research: As we incorporate the critical stance toward the ideological force of scientific and technological discourse that cultural studies approaches entail, how can we account for the agency that individuals appear to exercise without reducing such agency to either the filling of preexisting subject positions or strategic use of language that allows an individual to transcend such predefined positions?

In response to this question, I have demonstrated that although the forms of resistance that interviewees described might be understood as the acting out of subject positions that preexist their actions, not all of the effects of these resistant acts are contained within the scripts that define such positions. In fact, some of the acts of resistance that interviewees described entail reading and writing their bodies into the disciplinary grid in ways that disrupt the sense of this grid. In these examples, it is not only literate practices but also physical actions and ways of using the body (especially those that threaten the social order by doing things deemed impossible or violating cultural norms) that become integral to understanding how individuals resist the authority of powerful institutional discourses.

These findings contribute to ongoing academic conversations about the ways in which an awareness of rhetorical agency and resistance distinguishes technical communication scholarship from similar research in fields such as cultural studies. But in a culture in which mothers are often assigned primary responsibility for their infants' health and well-being, regardless of whether the prevailing discourse at a given historical moment privileges breast- or bottle-feeding, conversations about rhetorical agency and resistance in the face of the ideological force of medical discourse have implications that extend beyond academic conversations. Just as placing too much emphasis on the ideological force of such discourses can obscure individuals' abilities to resist such force, placing too much emphasis on the ability to resist might lead us to ignore the dangers and difficulties people can encounter in disobeying medical authority or violating cultural norms. Such an emphasis might be especially risky if it fuels the tendency to assign women too much blame for the difficulties they can encounter in aspects of life such as infant feeding.

As technical communication scholars continue to incorporate cultural studies theories and methods into our investigations of the interfaces between individuals and scientific and technological discourses, we need to keep these political realities in mind. If we continue to resist the tendency to oversimplify in either direction, we can continue finding new and better ways to acknowledge rhetorical agency and resistance without using this awareness to blame individuals for complex social problems. In particular, this study has suggested, we need more studies of the long-term effects of the rhetorical agency and resistance that women enact against medical discourse on aspects of life such as infant feeding. Such research is important because the outcomes of resistance are always possibilities, not guarantees; just as sense can be disrupted through individual acts of resistance, it can be put back into place through new forms of disciplinary rhetoric that aim to further restrict what is possible and permissible. Until further research is done, we cannot know which of these outcomes will prevail in the case of resistance to the disciplinary rhetorics of breastfeeding.

REFERENCES

American Academy of Pediatrics, Work Group on Breastfeeding. (1997). Breastfeeding and the use of human milk. *Pediatrics, 100,* 1035–1039.

Apple, R. D. (1987). *Mothers and medicine: A social history of infant feeding, 1890–1950.* Madison: University of Wisconsin Press.

Biesecker, B. (1992). Michel Foucault and the question of rhetoric. *Philosophy and Rhetoric, 25,* 351–364.

Blum, L. M. (1999). *At the breast: Ideologies of breastfeeding and motherhood in the contemporary United States.* Boston: Beacon.

Britt, E. C. (2001). *Conceiving normalcy: Rhetoric, law, and the double binds of infertility.* Tuscaloosa: University of Alabama Press.

Burger King apologizes to breast-feeding mom. (2003). *CNN.com.* Retrieved from http://www.cnn.com/2003/US/West/11/12/breastfeeding.apology.ap/index.html

de Certeau, M. (1984). *The practice of everyday life* (S. Rendall, Trans.). Berkeley: University of California Press.

Foucault, M. (1977). *Discipline and punish: The birth of the prison* (A. Sheridan, Trans.). New York: Vintage.

Foucault, M. (1978). *The history of sexuality Volume I: An introduction* (R. Hurley, Trans.). New York: Vintage.

Hausman, B. L. (2003). *Mother's milk: Breastfeeding controversies in American culture.* New York: Routledge & Kegan Paul.

Herndl, C. (2004). The legacy of critique and the promise of practice. *Journal of Business and Technical Communication, 18,* 3–8.

Koerber, A. (2002). *U.S. breastfeeding education and promotion, 1978–99: A feminist rhetorical analysis.* Unpublished doctoral dissertation, University of Minnesota, St. Paul.

Maher, V. (Ed.). (1992). *The anthropology of breastfeeding: Natural law or social construct.* Oxford, England: Berg.

Quandt, S. A. (1995). Sociocultural aspects of the lactation process. In P. Stuart-Macadam & K. A. Dettwyler (Eds.), *Breastfeeding: Biocultural perspectives* (pp. 127–144). Hawthorne, NY: Aldine.

Reeves, C. (1996). Language, rhetoric, and AIDS: The attitudes and strategies of key AIDS medical scientists and physicians. *Written Communication, 13,* 130–157.

Schryer, C. F., Lingard, L., Spafford, M., & Garwood, K. (2003). Structure and agency in medical case presentations. In C. Bazerman & D. Russell (Eds.), *Writing selves/writing societies* (pp. 62–96). Retrieved March 20, 2004, from http://wac.colostate.edu/books/selves_societies

Scott, J. B. (2003). *Risky rhetoric: AIDS and the cultural practices of HIV testing.* Carbondale: Southern Illinois University Press.

Spivak, G. C. (1992). More on power/knowledge. In T. E. Wartenberg (Ed.), *Rethinking power* (pp. 149–177). New York: State University of New York Press.

Van Esterik, P. (1995). The politics of breastfeeding: An advocacy perspective. In P. Stuart-Macadam & K. A. Dettwyler (Eds.), *Breastfeeding: Biocultural perspectives* (pp.145–166). Hawthorne, NY: Aldine.

Yalom, M. (1997). *A history of the breast.* New York: Knopf.

Amy Koerber is an assistant professor in technical communication and rhetoric at Texas Tech University. She has previously published articles in the *Journal of Business and Technical Communication* and *Women's Studies in Communication.*

REVIEW

Tracy Bridgeford, University of Nebraska at Omaha, Editor

Power and Legitimacy in Technical Communication, Volume II: Strategies for Professional Status. **Edited by Teresa Kynell-Hunt and Gerald Savage. Amityville, NY: Baywood, 2004. 219 pp.**

Reviewed by Michael J. Salvo
Purdue University

In *Power and Legitimacy in Technical Communication, Volume II: Strategies for Professional Status*, Kynell-Hunt and Savage continue tracing the centripetal and centrifugal forces that are simultaneously coalescing and dispersing power and legitimacy in the field. My review (Salvo, 2004) of Volume I concluded that the

> text engages the discourse of power/authority without offering false choices or easy solutions, nor does it assert a single or final definition of the field, discipline, specialization, or profession, for the text privileges none of these titles, neither willing to believe nor be comfortable in its unbelief. (p. 237)

This second volume articulates similar tensions as it traces pathways to power and legitimacy, offering both practitioners and academics numerous potential constructions of the future of the field. Although the first volume works to avoid privileging any strategy, many essays in this volume embrace the difficulties and ambiguities of postindustrial, global capitalism and postmodern culture as a source for rhetorical theory building, concluding that we may never find ourselves on solid ground and so we need to learn to sail.

The book consists of three parts. The first offers historical grounding for building strategies for attaining professional status, presenting three historical essays that promise to situate current and future attempts to empower practitioners and academics in technical communication. The second part offers approaches for rethinking contemporary practice that is aimed at increasing the cultural and workplace authority of communicators. The third part, titled "Strategies for Alternative Futures," focuses on long-term imaginings of the field and rearticulates strategies for developing technical communication in the long term. Without being constrained by current limitations, later essays in the book describe the futures we can or should strive for and how the field can be strategically developed to take advantage of these future contexts.

Part 1 offers three essays under the subtitle "Historical Perspectives for Present and Future Strategies." The first essay, a reprint of Teresa Kynell-Hunt's (1999) "Technical Communication from 1850–1950: Where Have We Been?" first appeared in *Technical Communication Quarterly*. It remains a timely and effective introduction to "the curricular shifts that led to the formation of a recognizable technical communication pedagogy" (p. 12). Kynell-Hunt's history is particularly interesting as it articulates events prior to Connors's (1982) "Rise of Technical Writing Instruction in America," which is not reproduced here but in Johnson-Eilola and Selber's (2004) collection, *Central Works in Technical Communication*. I expected the next essay to complete Kynell-Hunt's historical narrative; however, Elizabeth Tebeaux offers a jeremiad on the culture wars.

Contrary to Tebeaux's fears, the field does not invent or import unnecessarily complex theories. Rather—and the rest of the collection attests to this fact—technical communication creates theories with which to model, investigate, and comprehend an increasingly slippery and complex world, a complexity that Tebeaux accepts as real. It is not theory that makes the world unnecessarily complex; rather, it is the difficult work of engaging a complex world that makes theory-building a necessity. As other essays in this collection illustrate, carefully developed, meticulously applied, and hard-won knowledge emerging from the cycle of application, theory building, reapplication, redesign, retheorizing, and so on, is part of the strategic development of effective technical communication that builds solutions for teaching postindustrial rhetors.

Elizabeth Overman Smith provides an effective literature review that traces the "transmit" of the field, the "common terms, metaphors, and models for the problem, the methods for exploring the problem, and the practical application of the concepts and procedures to the problem situation" (p. 51). It is an effective introduction to the literature for newcomers to the field, and it contextualizes the collection's speculative contents.

In the second section, Marjorie Davis, Louise Rehling, and Robert Johnson offer different and distinct visions of the field. Davis effectively and convincingly argues for professional status and concludes that such status will become available only through codification of content, formalization of curriculum, and standardization of credentials. The argument is convincing until the reader realizes that the benefits of traditional professionalism are eroding in the postindustrial age.

Rehling offers a surprising account of the practitioner–academic debate that finds problems in the misunderstanding and poor definition of roles of these two less-than-distinct groups. The problem, as I interpret Rehling's argument, is not the oft-heard criticism that academics are out of touch with the needs of the workplace. Rather, Rehling's critique is that communication has been one-sided, from the workplace to the academy, and that a better balance needs to be struck. This rings true with my workplace-consulting experience in which practitioners wanted more than descriptive commentary on what they were doing. These clients wanted a vision of what else could be done, of medium- and long-range goals that would improve both their status and work satisfaction, and perhaps even some insight into the global marketplace in which they compete. This vision is consistent with Rehling's "two-way reality" in which the academy and workplace present "mutual influence and exchange" (pp. 92–94). The essay ends with concrete suggestions for supporting effective communication and collaboration between workplace and campus.

Johnson's essay "(Deeply) Sustainable Programs, Sustainable Cultures, Sustainable Selves" is a gem but an odd fit in this section, which is devoted to "Strategies for Contemporary Practice," and seemingly is a better fit for the final section. Yet its placement does not detract from the power of its argument. The essay lucidly offers three strategies for realizing power and authority as technical communicators:

- Redefine growth as it applies to technical communication programs.
- Become stewards of the technologies that fall within the purview of technical communication.
- Foster a sense of technical communication's responsibilities outside our immediate academic and workplace contexts. (pp. 116–118)

Johnson's second suggestion strikes me as further development of the author's previous articulations of *technical rhetors*. He suggests shifting from reaction to technological change on the one hand, to participation in that change on the other, asserting that technical communicators must stop seeing themselves as subject to technological shifts. By participating in the production and development of technologies important to production and communication, technical communicators become more than technicians; we become caretakers not just of *techne*, but we also assert *praxis*. More than any other call for professionalization, such critical self-awareness is necessary for improving the authority of technical communicators whether on campus or on the job.

The four essays that make up the final section offer long-term strategies for building workplaces and professional practices that are, first, hospitable to technical communicators, and by extension, more humane workplaces, more ethical workplace cultures, and more nurturing corporate, academic, nonprofit, and public institutions. It is important to remember that these essays aim to reverse trends that have narrowed opportunities for rhetorical intervention in technological and workplace design and development. They suggest what technical communication can become while remaining consistent with the field's origins and histories.

Rude's essay "Toward a Definition of Best Practices in Policy Discourse" offers a hopeful vision and optimistic role for technical communicators to play in the construction and communication of cultural attitudes and social values even if, as the essay records, recent experience has gone contrary to its vision.

> The story of bureaucracy dominating or excluding citizens occurs too regularly to ig-
> nore. But other cases, mostly local and small, offer some insight into the practices
> that enable citizens to participate in constructing policies that honor and support hu-
> manity and the democratic process. Disciplinary knowledge is available to improve
> practice. (pp. 139–140)

Rude's essay defines technical communication beyond the narrow confines of instruction sets and information technology as it moves the field's focus from functional texts to an enriched and complicated set of rhetorical concerns aimed at changing cultural realities. Such strategy builds on Johnson's essay and moves readers toward reproducing the conditions of (rhetorical) production—away from reactive tactics and toward developing proactive strategy. I hope Rude develops this idea further in subsequent publications.

Roundy Blyler's essay "Critical Interpretive Research in Technical Communication" offers a critical–theoretical vision of technical communication's future. The essay argues that theory and critique are necessary to effectively account for contextual and discursive flows of power in the professional workplace, and, indeed, it is often academics who remind practitioners of any profession's social responsibility. Blyler explicitly invokes cultural studies as an important analytic tool for technical communication practitioners that may provide insight into discourse in organizations.

Savage, offering a satisfying intellectual close to the two-volume collection, asserts that "modernist notions of profession are increasingly inappropriate to the theory we are developing and, more significantly, inappropriate to the circumstances and needs of our practice" (p. 176).

Here, in the penultimate essay of the collection, is one of the editor's statements regarding the developing discourse of power and authority. If we follow advice like that offered in the essays of Tebeaux, Smith, or Davis, we would have a potent and powerful profession constructed for an age that has passed, for circumstances that no longer abide, and for exigencies that have lost their discourse-constructing power. We would have, as Savage indicates, a perfectly conceived modern, industrial, national profession that resides, unfortunately, in a postmodern academy, a postindustrial workplace, and a globalized world. Technical communicators, envisioned as techno-rhetorical tricksters, fools, and sophists, invent opportunities for cultural power and authority such as the visions offered by Johnson, Rude, and Blyler. Not content to accept limited roles defined by powerful interests, Savage describes the technical communicator as one who builds authority and enables action by crossing boundaries, asserts the value of citizens, users, and audience members, and, "like the trickster, is an agent of social change" (p. 183).

As constructed by Savage, technical communicators are positioned and prepared to help organizations shift from the concerns of modern industry and its reliance on expert solutions to postindustrial knowledge work. Savage argues that such a role is "useful and disruptive," "rewarding and risky," and that "circumstances tend to consign technical communicators' practice to the periphery of the corporate cultures in which they work" (p. 189). The postmodern role Savage describes is difficult, thankless, and marginalized, but wholly worth pursuing, having more in common with the ancient concept of *metis*, which Savage aligns with navigation as a contextualized way of knowing that literally takes advantage of the winds of change.

The final essay is a reprint of M. Jimmie Killingsworth's (1999) "Technical Communication in the 21st Century: Where Are We Going?" originally published in *Technical Communication Quarterly*. In the essay, he speculates on the pedagogical value of using science fiction, or SciFi, as a way into theory for resistant students and teachers. Killingsworth offers an interesting pedagogical suggestion but simultaneously reveals the wide gap between the rich complexity of the field's developing theory and apparent limits to what can be taught.

Kynell-Hunt and Savage's second volume of *Power and Legitimacy in Technical Communication* accurately presents tensions inherent in any attempt to professionalize the field. Indeed, as I read and reread the book and began drafting this review, I was very much aware of the tensions present both in the field of technical communication and the competing visions of the future as presented in this book. The collection ultimately does not deliver what the subtitle promises, that is, strategies for becoming professional. Instead this collection of essays offers readers something more complex and perhaps more valuable: Sustained, multivocal inquiry into the meaning, promise, and limits of professionalism in a time of change. Although it does not present a singular vision of the postmodern professional, it instead describes the postindustrial workplace without flinching from the challenge change brings, and it even begins to describe some of the potential rewards of change. The first waves of digital technology and network culture brought with it the metaphor of surfing. Savage closes the collection with reference to *metis*, to the skill necessary to sail. The collection implies that rather than deciding whether we are modern or postmodern, industrial or postindustrial, we need to learn to harness the energy of these discussions, to ride the energy of change itself, and to learn to sail across stormy seas. The challenge is to devise curricula, describe careers, and make knowledge that informs and prepares for this brave new fluid world.

REFERENCES

Connors, R. J. (1982). The rise of technical writing in America. *Journal of Technical Writing and Communication, 12,* 329–352.

Johnson-Eilola, J., & Selber, S. A. (2004). *Central works in technical communication.* New York: Oxford University Press.

Killingsworth, M. J. (1999). Technical communication in the 21st century: Where are we going? *Technical Communication Quarterly, 8,* 165–174.

Kynell-Hunt, T. (1999). Technical communication from 1850–1950: Where have we been? *Technical Communication Quarterly, 8,* 143–151.

Salvo, M. J. (2004). Power and legitimacy in technical communication, volume I: The historical and contemporary struggle for professional status [Book review]. *Technical Communication Quarterly, 13,* 235–237.

CALL FOR PAPERS

Technical Communication Quarterly, **Special Issue 16.3**
Technical Communication in the Age of Distributed Work

Shoshana Zuboff and James Maxmin are excited about it and see it as a moment of new liberation and choice for consumers and workers alike. Gilles Deleuze saw it as horrifying, even worse than the disciplinary society Michel Foucault described. It goes by many names: distributed capitalism, the control society, the informatics of domination, the support economy. Whatever its name, the characteristics are the same: Control over organizations is as distributed as ownership is in managerial capitalism; digital technologies play a vital enabling role; consumption is individuated, taking the form of the desire for unique identities and unique experiences; direct relationships between customers and businesses become more important; and customers look for stable beneficial relationships among consumers and producers that support these individual experiences. These needs are supplied not by large, vertically integrated companies but by temporary "federations" of suppliers for each individual transaction. These federations are endlessly recombinant. Work is fragmented temporally, geographically, and disciplinarily. Lifelong employment is replaced by what Zuboff and Maxmin call "lifelong learning"—what Donna Haraway calls continual deskilling and retraining.

We can see the early signs of distributed work in the service sector, in the outsourcing of technical support, and in places like eBay and Craig's List. But we can also see it in the rise of homeschooling, the weakening of unions, the shift from stable identity politics to unstable subsegments, and the popularity of automobile customization. We can detect it in the proliferation of time-management methods, the popularity of distance education, the increasing importance of content management systems, and the early success of Howard Dean's campaign. We can trace its contours in Brenton Faber's discussion of corporate universities; Johndan Johnson-Eilola's explorations of dataclouds; and Teresa Harrison and James Zappen's development of online community spaces and attendant research methods.

What does distributed work mean to us as technical communicators? How is it changing our field? Should we adapt to it, critique it, or resist it?

In this special issue of *Technical Communication Quarterly*, we will discuss distributed work's implications for technical communication theory, methodology, pedagogy, ethics, and practice. In particular, we will consider topics such as:

- How is technical communication practice changing, and how will it change in the future, as it adapts to distributed work? How will it accommodate, resist, or redirect?
- How do we teach technical communicators who expect to go into the support economy? What are our political–ethical responsibilities and our logistical challenges? What changes do we need to make to pedagogical theory?
- What roles will technology play in an economic climate in which knowledge, expertise, and intelligence are widely distributed? For instance, how can software documentation survive when users routinely Google for answers?
- What theoretical frameworks are useful for theorizing the shift to distributed work? What case studies can be used to illustrate it and explore its implications for technical communication?
- What research methods do we need to adapt or develop to apply to distributed work in technical communication research? What methods should we abandon?
- Finally, what are the contours of distributed work? What are its promises and horrors?

Schedule

- 1- to 2-page proposal for paper: March 15, 2006
- Full paper (if proposal is accepted): June 30, 2006
- Scheduled publication of issue: Summer 2007

Contact information: Send proposals in .DOC, .RTF, or .HTML to clay.spinuzzi@mail.utexas.edu.

Also, please contact the editor by e-mail if you would like to be considered a reviewer for this special issue.